# Basics of Speed Mathematics

Dedicated to my team

astrarka
DIFFERENTIATED LEARNING
PLATFORMS FOR CHILDREN

ISBN: 1453798668
EAN-13: 9781453798669
First Edition 2010

# Table of Contents

# Foreword

Several books have been written on the subject of Speed Mathematics, Vedic Mathematics and other systems of numerical manipulation. This book is different – in that it brings together the techniques from Vedic Mathematics and Trachtenberg System together. We have not attempted to do a comparative study of these techniques and make a judgment on which one is better. Instead we have simply, presented the techniques in a sequence that makes most sense. There are inherent strengths in these two approaches. While Vedic System has very short and pithy statements, which encapsulate the essence of a technique [base multiplication is a fine example of this], the Trachtenberg System offers valuable insights into how very simple adjustments in the way we do things, go a long way in improving accuracy. The treatment of speed addition and preliminary treatment of conventional elementary school division is an excellent insight. All systems of knowledge rely on profound analogies. The argumental division is an excellent case in point. The section underscores an important point that a polynomial in $x$, is simply a decimal number when the value of $x$ is equal to ten. This equivalence drives home a very useful insight. We can now draw several algebraic techniques into play while handling numbers, and similarly, we can use number manipulation techniques while we handle problems in Algebra. At the core of education, is the need of conceptual understanding. The good demonstration of our conceptual understanding of a subject is related to our ability to draw these analogies and make such meaningful connects.

We sincerely hope that the student is able to get a good grasp of the subject and the techniques after working with the content of this book.

Astrarka Educational Solutions Private Limited.
Bangalore, India.

# Preface

In June 2009, we started to work on a video of Basics of Speed Mathematics. The motivation was very simple. We wanted to offer the student a bouquet of techniques for flawless computation of problems related addition, subtraction, multiplication and division. Once the student is familiar with the underlying principles of numerical manipulation, he would be able to make a decision on what works well for him given a certain problem in Arithmetic.

Furthermore, several schools and institutes openly focused on speed and speed alone. We beg to differ. Speed at the expense of accuracy is useless. There is really no point in racing through a highway at a blistering pace if there is no guarantee of safety. It is therefore very clear that any school of thought that does not focus on inherent error checking techniques is simply not comprehensive enough. So, we decided to spend a section of this work on error detection techniques using digital roots.

One other area of focus has been speed addition techniques. The method you will see appears very elaborate, but in fact, it is robust and complete. Most books on speed math that we have come across do not talk of these methods. The Trachtenberg system is a place where we looked for these techniques.

Multiplication is the crown jewel of speed mathematics. We have devoted a good chunk of material on this topic. We have presented several key techniques from the Vedic and Trachtenberg school of thought. Both these schools offer powerful insights into Speed Math. It would be unfortunate if we fall prey to the temptation of picking one over the other as the golden standard.

Several man months have been invested on working through the basic material in preparation of the speed mathematics video. The idea to convert this into a book was an afterthought based on feedback from parents, who believed that they really needed a workbook of sorts to help their children with these problem solving techniques.

The feedback from the parents and teachers, who championed our video products, laid the foundation stone for this book. It has been a humbling experience, although, we must add – that discovering the child in us and going through the materials of elementary school Mathematics filled our work days with immense joy.

It would be impossible for us to acknowledge all the people that have contributed to this mammoth effort. This is the problem related to completeness in our enumeration. However, it would be unfair if we did not thank a few people whose contributions stood out during the design, production and review process of the video[1] and the book. Monica Ranjit helped us out with the content of the Trachtenberg System, and Santosh Govindarajan bore the brunt as he created and solved well over a 1000 problems in preparation of the video shoot. R Balasubramanian was always ready to help with smart little perl routines for generating random numbers while building the exercise problem set. Without their colossal effort, this book would have been an exercise in futility.

We would like to thank all the staff of Astrarka for their contributions, support and assistance throughout this project.

August 2010
Chandramouli Mahadevan
Bangalore

---

[1] The 25 hour video on Basics of Speed Mathematics is a 6 DVD set produced by Astrarka. For additional details about the product and its features, please feel free to contact sales@astrarka.com or http://www.astrarka.com

# 1 Introduction

There was a time in ancient India when Mathematics, and perhaps even education was imparted using Sanskrit Sutras. Sutras are aphorisms. These are short sentences – usually a few words or sentences long, which contained a specific formula or way of solving problems. Our attempt in this book is very simple. We are not going to sing praises of the wisdom of our ancestors. We will understand and apply these aphorisms to our day-to-day Math problems. This will help us become better equipped to handle complex Math as we progress through our schooling phase. Therefore, this is not a comparative study of techniques for Speed Math. Our attempt is to teach you develop basic Math skills – thus help you understand and enjoy the language of numbers; identify patterns and handle the quirks and manipulation of numbers with ease. With practice, you will start experiencing the fact that the anxiety associated with learning Math gradually dissolves, making way for a sense of appreciation and even command over the subject. Constructive curiosity will thus lead you to become confident in dealing with the subject.

## 1.1 Review of learning objectives

Speed Mathematics deals with a collection of techniques. If you master these techniques, you will dramatically improve your speed and accuracy in Math. We will also emphasize on techniques to check for accuracy after a solution reached. This step will help you identify errors and give you an opportunity to correct you solutions. Therefore, we have three major learning objectives.

1. Techniques for handling multiplication, division, addition and subtraction.
2. Techniques for ensure that the techniques when applied do produce the desired results.
3. Techniques to check for accuracy of solutions.

For doing this, we will have to commence with an introduction of terms and phrases which we will frequently use. These terms and definitions form the basic building blocks of our language of speed math.

For some reason, most societies relate ability to solve problems to intelligence. People with an ability to solve problems fast have always been considered very intelligent. Those with an uncanny ability to solve problems not only quickly, but also accurately, are considered super intelligent. Is this ability a blessing or a skill? We think this is a skill and each and every kid can develop this skill if he or she is willing to invest time and effort to learn, understand and practice these techniques. To be good at anything takes practice. It takes of hours and hours of effort and practice, for someone to be an expert like say, Roger Federer or Raphael Nadal. These men have dedicated several hours every day to practice their art. To be good at Math requires a similar attitude. We must sacrifice boredom; practice the techniques by solving problems. Let us now capture these points into a basic expectation from each of you. Let us make a personal commitment to work towards becoming very good at number manipulation.

## 1.2   Basic Expectations

These are simple rules of thumb for getting better at the art of speed mathematics.

1.   Always have a paper and pencil handy with you while going through this material / class. I would recommend that you set aside a notebook for taking notes, solving problems and practicing what you learnt in a certain section.
2.   Practice by solving problems and then by solving even more problem. The more you practice the better you become.
3.   Believe in yourself. You are good at Math; and if you don't think so – believe in your ability to become good at Math. Math is not hard; like Music, it takes practice and dedication.
4.   Because you have learnt do things in a certain way at school, it will take a while for you to get used to doing this differently. Be reminded that all the methods, of computation, work. What we are going to learn here is a different way to look at the problem solving process.

At the end of the course, we will look at formal proofs for our techniques. We strongly recommend a bright beginner to keep away from this and invest his energy on the learning aspect of the course.

# 2   Building Blocks for Speed Math

This is a textbook and a workbook for students. We are going to deal with Speed Math techniques. Speaking of Speed Math there are multiple schools of thought that exist in this domain; there is a Trachtenberg System from Europe, there are Chinese speed math techniques based on Abacus, and there are many more. Perhaps it is safe to say that every civilization had a math technique; some of these techniques are available today. This course is not a comparative study of various methods for Speed Math; rather this is simply a way of exposing the techniques and giving you practice in using those techniques in day-to-day math problems. In doing so it is our hope that the anxiety and fear associated with conquering math will gradually vanish leading to greater self confidence in taking on challenges in math and education at large.

For some reason, most societies relate ability to solve problems to intelligence. People with an ability to solve problems fast have always been considered very intelligent. Those with an uncanny ability to solve problems not only quickly, but also accurately, are considered super intelligent. Is this ability a blessing or a skill? We believe that this is a skill. Therefore it can be developed. Each and every kid can develop this skill if he or she is willing to invest time and effort to learn, understand and practice these techniques.

It is safe to say that in Ancient India, mathematics and perhaps even education was handed out through aphorisms called Sutras. In the pre-printing press era, much of knowledge was captured as aphorisms the world over. Typically they were a couple of lines long and they had all the knowledge required to perform a certain step or an equation. This course is not about praising or analyzing the aphorisms in detail. We will simply extract, distill the essence of the aphorisms and offer those techniques in a form that is easy for us to grasp and internalize and use in our day-to- day problems. This will help us become better equipped to handle complex Math as we progress through our schooling phase. Therefore, this is not even a comparative study of techniques for Speed Math. Our attempt is to teach you develop basic Math skills – thus helping you understand and enjoy the language of numbers; identify patterns and handle the quirks and manipulation of numbers with ease. With practice, you will start experiencing the fact that the anxiety associated with learning Math gradually dissolves, making way for a sense of appreciation and even command over the subject. Constructive curiosity will thus lead you to become confident about the subject.

You need to have a paper and pencil at all times while you go through the material in this book. The basic expectation is that you will solve the examples and other problems; write them down in your own way and appreciate the patterns and techniques. This is not a book, which you can read casually and say, 'Wow, how magical.' That's not the intent. The intent is for you to become an expert, the intent is for you to have the self-belief that by going through this book you can be very, very good at speed math; you can be fast, you can be accurate. That is the main purpose of this book. You don't have to be neat; it does not have to be a work of art. Just scribble along of that is how you learn new concepts and practice them. The idea is to try out these mental techniques and in doing so, we will become very good at speed math.

Let us take a look at how mathematics is taught. We are taught techniques to solve problems. In doing so it is only human that we make mistakes. These errors cost us a few points in our tests, exams and assignments. The ways we are encouraged to correct or minimize these errors by revise our answers one more time; or going through the answer sheet again or solve the problems one more. There is a limited value in this feedback because by repeating the same steps again you are most likely to commit the same errors too. Of course, if you are fortunate you can detect an error and things may be fine for you. But, is this the way the process of checking out the accuracy of a solution should work? Not at all. Therefore in addition to actually focusing on **speed** math, we would like to focus on a very important concept of checking for accuracy. In a recent school function I heard the principal proudly announce, '*We have a course on Speed Math where the sole goal is to be fast at doing things*'. Students are encouraged to focus on improving the speed of problem solving. Speed at the expense of accuracy is meaningless. You need to be accurate and quick as well. In doing so, the level of accuracy and perfection that you attain in the process will be enhanced and your self-confidence will gradually start growing. Your belief in your ability to conquer math will dramatically improve and that is the whole purpose of actually learning these skills. So, believe in yourself, pick up a paper and pencil and we are good to go.

As we go through the techniques, we can wonder if these techniques are magical. Are these founded on formal proofs? Is there a basis for saying that these techniques work in all situations? So a curious mind is bound to ask this question, *"Why should I believe that these techniques work across the board?"* For helping such children, we will spend a few sections as applicable in actually presenting algebraic proof for the proofs presented, for the techniques presented so you know why is it that things work the way they do. It is like lifting the car's bonnet and looking under the hood and saying, 'Ah, I think I understand', and sometimes understanding how things work helps improve our own understanding of the technique itself and we get more and more comfortable in using those techniques. But for a beginner, I would recommend that you skip these sections. If you just want to go through it, listen through it, then that is fine. If you don't understand it, spend some more time at math and you will eventually start understanding the patterns.

Solving math problems is pretty much like handling chess. Understanding the basic numbers, symbols is easy, just like understanding how a pawn moves, how a queen moves, how a knight moves on the chess board, with a little practice you can play a game of chess. But to be good at chess you need to be able to extract the pattern on the board and react to the pattern. In other words our confidence in math is directly proportional to how comfortable we are with dealing with numbers, dealing with number patterns and manipulating those number patterns. So, we require a mechanism to start figuring out how to fall in love with these number patterns, how do we identify these patterns, how do these patterns emerge and pop out of the paper as you lay them down on a piece of paper. Once you start understanding how these patterns emerge, you would realize that your process of problem solving would improve. And this will come with practice. To be good at anything takes practice. It takes of hours and hours of effort and practice, for someone to be an expert like say, Roger Federer or Raphael Nadal. These men have dedicated several hours every day to practice their art. To be good at math requires a similar attitude. We must sacrifice boredom; practice the techniques by solving problems.

## 2.1   Introduction to the World of Numbers

We are going to commence our journey into the world of numbers. There will be a method in our madness. We will solve problems. We will solve lots of problems. While solving the problems, we will look at the patterns they exhibit. We will continue to apply these patterns until they break down. As the techniques fall short while dealing with a certain scenario, we will expand or extend the same techniques – in order to take care of the situation. This process of generalization will help us to develop the art of looking for and looking at patterns. This is the heart of the Speed Math program.

Typically, a verbal definition of a concept sounds lot more complicated when you think about it in English. They all look like tongue twisters and some of them are confusing. Every time you come across such statements, do not be overwhelmed. Focus on the fundamental patterns and principles. Solve problems using these fundamental concepts. Understand the intent that the definition or the statement is trying to communicate. Understanding **that** is more critical than knowing a verbatim definition of a technique or a rule.

First of all, math predominantly revolves around ten symbols - 0, 1, 2, 3, 4, 5, 6, 7, 8, and 9. Most problems in arithmetic can be handled within these symbols. Of course, for decimals you need a dot. A few more symbols like a slash for a fraction and so on. But predominantly the number manipulation relies on these ten symbols.

One of the things that we do with these ten symbols is this. You can represent a number, no matter how large or how small, with a combination of these ten symbols. For doing this, we use two significant concepts. This is the concept of face value and place-value. Some of you may be familiar with this concept. But bear with me, because we are kind of now opening a box of Lego pieces and laying them out on the table. In doing so we're going to deal with the building blocks of this manipulation. Therefore it is important for us to revisit the fundamentals.

A number is represented by a string of digits. Let us start with an example. Consider the number 261. The number 261 stands for two hundred and sixty one. This number consists of two Hundreds, six Tens and one Unit. In some books instead of unit they use a term called Ones. So it is the same as two Hundreds, six Tens and one Ones.

Similarly, 6261 stands for six thousand two hundred and sixty one. Therefore, 6261 consists of six Thousands, two Hundreds, six Tens and one Unit. The concept of face value of a number is simple. It is simply the value of the number. There are 10 symbols with a face value. 1 is worth one Unit, 2 is worth two Units and 3 is worth 3 Units. 9 is worth 9 Units.

Face value of a number is the simply what that symbol is worth irrespective of where it occurs in a number.

In the decimal number system, we also have a concept of place value. Each digit-position from right is left increases 10 fold. The right most place of a number is called the Units place. It is worth 1. The position to the left of it is the Tens place. It holds the number of tens in the number. The position to its left is the Thousands place, which tells us the number of thousands in the number.

| Millions | | | Thousands | | | | | |
|----------|------|-------|----------|------|-------|----------|------|-------|
| Hundreds | Tens | Units | Hundreds | Tens | Units | Hundreds | Tens | Units |
| 6 | 4 | 2 | 1 | 4 | 6 | 5 | 8 | 9 |

Fig 1.1: Basic ways of reading a number off a place value chart.

*We read this number as "Six hundred and forty two million, one hundred and forty six thousands and five hundred and eighty nine". The face value of 6 remains the same. However its place value is different. The 6 in the leftmost digit is worth 100-million while the other six is worth a thousand, because it occurs in the Thousands place.*

| 100 millions | 10 millions | millions | 100,000s | 10,000s | 1000s | hundreds | tens | units |
|--------------|-------------|----------|----------|---------|-------|----------|------|-------|

Fig 1.2: The American Place Value System

*Note that the place value to the left of ten-thousands is called hundred thousands, millions, 10 million, 100 million, billion and so on. Contrast this with the Indian place value system shown below, where we use terms like lakh [ hundred thousand ] and a crore [ ten million ]. While terms and terminologies vary, the basic underlying structure of the decimal number system remains the same.*

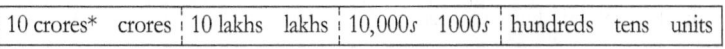

| 10 crores* | crores | 10 lakhs | lakhs | 10,000s | 1000s | hundreds | tens | units |
|------------|--------|----------|-------|---------|-------|----------|------|-------|

Fig 1.3: The Indian Place Value System

*For a long time, we used to wonder if there are place values larger than a crore in this system. http://en.wikipedia.org/wiki/Indian_numbering_system gives us insights into larger place values like arab, kharab, neel, padma and shankh. But these are very rarely used and even less understood by millions in India.*

Key concepts are as follows:

- The place value of the right-most digit is Unit.
- The place value increases 10 folds in the each of the subsequent digits to the left of the Units place.
- The face value of the number is simply the numerical value of the symbol. It does not change with the position of the symbol in the number.

Let us check our understanding of these concepts. You will have to capture the place value and face value of six in each of the following numbers.

| # | Number | Place-value of 6 | Face-value of 6 |
|---|--------|------------------|-----------------|
| 1 | 6      |                  |                 |
| 2 | 16     |                  |                 |
| 3 | 161    |                  |                 |
| 4 | 671    |                  |                 |
| 5 | 6193   |                  |                 |

Fig 1.4: Exercise on place-value and face-value of numbers

## 2.2 Patterns

Numbers exhibit patterns. The idea is to look at a collection of numbers, usually in a series and infer the subsequent numbers in that series. We will start off with easy ones and continue to explore additional examples. As we gain an understanding of these patterns and series, we will start looking for patterns every time we solve problems. In most standardized tests, pattern-matching problems are still a useful tool in determining general aptitude levels of the participants.

The first pattern we are speaking about is 1, 2, 3, 4, 5 and the challenge in front of us is to identify what are the three numbers that come after the five. One look tells us that these are the first five numbers starting from 1 on the number line. Therefore, the next three numbers must be six, seven and eight.

The next series we will take a look at is 1, 3, 5, 7, and 9. Starting with 1, these are the first five odd numbers on the number line. Therefore, the next three odd numbers after 9 should be the ones that follow. These are 11, 13 and 15. Similarly, if you looked at the series 2, 4, 6, 8, and 10; it is clear that these are the first 5 even numbers starting with 2. Therefore the next three numbers that follow in the series must be 12, 14 and 16.

Let us take a look at the series 1, 4, 7, 10 and 13. To identify the pattern, you now need to identify the pattern or the relationship between successive numbers. In this series, we can see that each number is three more than its predecessor. Therefore the next three numbers in the series must be 16, 19 and 22. In the previous examples, where we identified the series as even numbers and odd numbers, we were doing the same thing. The difference between the successive numbers in the series was 2. We could have identified this pattern and we would have arrived at the correct answer. The point is that we can choose a pattern that is valid as long as that pattern is true for all numbers in the series.

| # | Series | Next three numbers |
|---|--------|--------------------|
| 1 | 5,10,15,20,25 | |
| 2 | 100,90,80,70,60 | |
| 3 | 0,1,1,2,3,5,8 | |
| 4 | 1,2,4,8,6,32 | |
| 5 | 5,75,10,65,15,55,20 | |

Fig 1.5: Identify the next three digits in this series

In fact, it is not necessary for the numbers to be laid out in the series. You can spread them spatially and look for patterns as well.

```
                    1
              2     3     4
        5     6     7     8     9
  10    11    12    13    14    15    16
```

In the triangle of numbers, we can see that the last number in each row is the square of successive numbers starting with 1 on the top. Therefore the next row must end with 25, which is five-squared. So by laying out numbers spatially like a triangle, we can come up with different interesting patterns. There is a whole topic in Mathematics related to triangular numbers that deals with such interesting patterns. The idea is that numbers do exhibit several interesting properties and patterns. As you gain some practice in looking for such patterns and designs - appreciating and extracting these patterns, you will see that your expertise in handling numbers is greatly improved. You will start "*seeing*" the underlying language of numbers. Familiarity, they say, breeds liking.

## 2.3   Basic Strategy

Our early counting experiences revolve around considering objects like candies, cookies or apples for counting. A typical problem simply asks us to count 5 apples and 3 apples – asking us to determine the total number of apples. Children use their fingers to go through the counting process. The primary idea is as follows. So long as the child can count using his / her finger to count, the basic operation can be successfully performed. In other words, a problem that can be reduced to determining the sum of five and three will yield a response as follows. Counting commences with one, two three, four and five – on one hand. Five fingers are open. Counting continues one the other hand -with one, two and three. Now three more fingers from the other hand are opened. Now the child simply counts all the open fingers – one, two three, four, five, six, seven and eight. Thus the final sum is eight. It is very important for us to understand the general thought process behind these problems and problem solving techniques.

1.   First, put all the numbers that you want to add in a column
2.   Add each one of them
3.   Complete the addition process
4.   The final result is the sum of the numbers

Eventually with practice, the child commences to recognize the sum of two numbers without going through the elaborate process. Some books refer to an addition table as shown in Fig 1.6.

| + | 1 | 2 | 3 | 4 | 5 | 6 | 7 | 8 | 9 |
|---|---|---|---|---|---|---|---|---|---|
| 1 | 2 | 3 | 4 | 5 | 6 | 7 | 8 | 9 | 10 |
| 2 | 3 | 4 | 5 | 6 | 7 | 8 | 9 | 10 | 11 |
| 3 | 4 | 5 | 6 | 7 | 8 | 9 | 10 | 11 | 12 |
| 4 | 5 | 6 | 7 | 8 | 9 | 10 | 11 | 12 | 13 |
| 5 | 6 | 7 | 8 | 9 | 10 | 11 | 12 | 13 | 14 |
| 6 | 7 | 8 | 9 | 10 | 11 | 12 | 13 | 14 | 15 |
| 7 | 8 | 9 | 10 | 11 | 12 | 13 | 14 | 15 | 16 |
| 8 | 9 | 10 | 11 | 12 | 13 | 14 | 15 | 16 | 17 |
| 9 | 10 | 11 | 12 | 13 | 14 | 15 | 16 | 17 | 18 |

Fig 1.6: An addition table. Each cell contains the sum of the column and row heads.

The technique that we will introduce now is slightly different. The idea is to start looking at numbers in terms of tens. We will begin with looking at various ways of getting to 10 by adding two numbers.

$10 = 9 + 1 = 8 + 2 = 7 + 3 = 6 + 4 = 5 + 5 = 4 + 6 = 3 + 7 = 2 + 8 = 1 + 9$

The last half is simply a rehash of the first. This is because $2 + 3 = 3 + 2$. When we are given a problem to solve, for example, twenty seven plus thirty two, we must build the ability to do the following mentally.

- $27 = 30 - 3$
- $32 = 30 + 2$
- $27 + 32 = 30 + 30 + (2 - 3) = 60 - 1 = 59$

The idea is to look at numbers as being deficient or in excess of some multiple of 10. This will means that the bulk of addition is simply counting up the number of tens. We can compute the excesses and deficiencies, which are usually small. This will complete our addition operation. We had made an observation before. When we start writing down the mental steps in English, things appear complicated. The goal for us is to get to a state where we look at 27 and what flashes in our mind is $30 - 3$; we look at 32 and what flashes in our mind is $30 + 2$; the rest is very simple and easy. So looking at each one of the individual numbers in terms of what is the deficiency of the number with reference to the nearest tens vastly improves the speed and accuracy for two reasons.

a. If the numbers are large; let us assume we want to compute 69 + 87; the classical way of addition would be as follows.

**Ex. 1:** 69 + 87
Solution:

$$
\begin{array}{r}
{}^{1}6\ 9 \\
+\ 8\ 7 \\
\hline
1\ 5\ 6
\end{array}
$$

In this method, we will have to deal with large numbers – 9 and 7; and then with a carry over. In the new method; we would look at 69 and 70 – 1 would flash in our mind; 87 would trigger 90 – 3; 70 and 90 would give 160; 160 – 4 = 156. In order for the mind to flash the right deficiency or excess when we see a number, we would need practice. Furthermore, the deficiency or the excess is typically much smaller in magnitude than the original number.

And we know that when we deal with smaller numbers, the chance of making an error diminishes significantly.

b. The other reason is that the entire operation of addition can now be performed mentally. We have eliminated the need for carry over as an explicit step. This is because we have split the original number into a multiple of 10 and a deficiency / excess. We can handle each of these things separately and merge them together in the final step.

The key to this method is to start looking at number in terms of "short of" or "over" a reference multiple of 10. Speaking of which, brings us to the next issue. Let us now consider 38. Both 30 + 8 and 40 – 2 are valid mental representations. Typically, we would choose 40 – 2 because 2 is a smaller number than 8 (30 + 8). And we want to always deal with small numbers. We want to eliminate the need for borrow or carry over operations. There are situations when we must look at 38 as 30 + 8; and we will get to that case in a short while from now.

Let us try manipulating and finding out under what circumstance what kind of numbers would you want to use. These are thumb rules. These techniques give you a flavor of how to approach, there is no hard and fast rule of saying under these circumstances do exactly this. It is kind of like riding a bicycle in a street, you have to react to the situation. You cannot have a road that says please go ahead and cycle and at the end of 20 feet apply brake. You will apply brake whenever you come across an obstacle or a vehicle or oncoming pedestrian. You will be careful not to run into a stationary or moving object. And that is what we mean by a rule of thumb. You look at the pattern and you want to react to the situation.

Let us look at another problem.

**Ex. 2:** 46 + 83

We could use the following steps mentally in order to arrive at the solution.
1. $46 = 40 + 6$
2. $83 = 80 + 3$
3. $46 + 83 = 120 + 9 = 129$

Or, we could use
1. $46 = 50 - 4$
2. $83 = 80 + 3$
3. $46 + 83 = 130 - 1 = 129$

Both these are valid worldviews. Typically, a sign flip can cause errors. It is a call you need to take.

The general thumb rule remains the same:

1. Look at numbers in terms of what those numbers are short off or above.
2. Start splitting numbers into tens and units
3. Deal with small numbers. Practice and learn to be comfortable with dealing with small numbers because probability of making errors with small numbers is vastly diminished.

## 2.3.1   Exercises for practice

**Prob. 1:** 28 + 7
**Prob. 2:** 78 + 8
**Prob. 3:** 69 + 4
**Prob. 4:** 36 + 9
**Prob. 5:** 87 + 8

If you start writing down each of these steps in a verbose fashion, the basic idea of mental math is defeated. The idea is for you to understand the essence of the technique, therefore you practice a situation and you are ready. It's pretty much like learning to ride a bicycle. When you drive around in your yard, you kind of understand where to turn, how to apply brakes, how not to fall. You perhaps even put a blockade here or there, an obstacle, a bucket, or whatever to learn to practice to maneuver around it. Eventually, that is all the simulation of the traffic. Similarly using these easy looking examples, we are actually trying to understand the underlying techniques.

Let us solve a simple puzzle. The rules are as follows.

Goal: Express 10 as a sum of three numbers.
- You cannot use 0.
- You can repeat numbers.
- You need to come up with unique combinations. This means 5+4+1 is the same as 1+4+5. They do not count as two solutions.
- Try to identify as many ways as possible in 1 minute

For example, 3+3+4 is one way; 5+4+1 is another.

This puzzle teaches you a few basic ideas. One of the three numbers cannot be 9. This will lead to 9+1+0 (incidentally this is a valid answer but...) and zero is not allowed. It will help you see what is allowed and what is not allowed. These sub-rules that you come up with are called constraints.

We can now make a few generic observations based on what we have done so far.

- We can reduce the complexity of addition by constantly looking for ways to get to a ten.
- This technique also eliminates the need for a carry-over to be handled explicitly.
- The excess or the deficiency is typically a small number. The chance of an error diminishes dramatically when we deal with small numbers.
- When we see a number or a set of numbers, we look for patterns to get a ten, before we start counting.

Some more examples will highlight the observations. They are easy because our focus is on the technique and we are trying to build a new skill of looking across a set of numbers and identifying the pattern – to get to a ten.

## 2.3.2 Exercises for practice

**Prob.** 1: 3 + 8 + 2
**Prob.** 2: 1 + 8 + 9
**Prob.** 3: 4 + 5 + 7 + 5
**Prob.** 4: 9 + 7 + 1
**Prob.** 5: 8 + 9 + 2 + 1
**Prob.** 6: 7 + 4 + 2 + 6
**Prob.** 7: 8 + 8 + 8 + 4 + 2
**Prob.** 8: 1 + 2 + 3 + 4 + 5
**Prob.** 9: 8 + 6 + 4 + 2

So if you were dealing with 2 digit numbers, the basic strategy does not change. If you are dealing with, let us say 33 + 28, you have to view this problem as two problems: 8 + 3; and 30 + 20. Once you decompose a larger problem into smaller manageable units, your speed and accuracy will improve. Before we do that, we would like to evolve a short hand for numerical representation.

63 = 6 tens and 3 units = 5 tens and 13 units = 4 tens and 23 units.

In normal math, every time we come across the number greater than 10 in a certain place value, we write down the units' part of that number in that place and carry over the remaining digits to the place value to the left.

We would like to use the concept of carry-over in a different way. We would like to retain all the carry-over until the last step and finally adjust the digits for the carry over step. We would call this step – "Collapse".

Let us look at an example.

**Ex.** 1: 19 + 9
Solution:

|   | 1 | 9 |   |
|---|---|---|---|
| + |   | 9 |   |
| | 1 | $_1$8 | We will subscript the 1 to indicate that this is a carryover digit |
| | 2 | 8 | We will collapse the carryover digit in the final step |

This means that we do not have to worry about the carry over and related adjustments after every single operation. Instead, we will simply write down the answer with a subscripted notation as shown above. We will handle all the carry-over digits in one go during the final step. This means that we can act like that there is no carry-over or borrow to worry about. No matter how large the number is we simply subscript everything except the units' digit of the number and moves forward. This simplifies the computation procedure and reduces the chance of error.

Here is another example to highlight the procedure.

**Ex. 2:** 89 + 77 + 62
**Solution:**

|   |   | 8 | 9 |   |
|---|---|---|---|---|
| + |   | 7 | 7 |   |
| + |   | 6 | 2 |   |
| |   | $_2$1 | $_1$8 | Carryover digits are subscripts |
| = | 2 | 2 | 8 | Collapse the carryover digits |

## 2.3.3 Exercises for practice

**Prob.** 1: 33+24+27+16
Solution:

```
      3  3
+     2  4
+     2  7
+     1  6
  _____
=  _____
```

**Prob.** 2: 24+17+28+16+3
Solution:

```
      2  4
+     1  7
+     2  8
+     1  6
+        3
  _____
=  _____
```

**Prob.** 3: 33+38+39+31
Solution:

```
      3  3
+     3  8
+     3  9
+     3  1
  _____
=  _____
```

**Prob.** 4: 15+13+27
Solution:

```
      1  5
+     1  3
+     2  7
  _____
=  _____
```

**Prob.** 5: 16+13+87+42
Solution:

```
      1  6
+     1  3
+     8  7
+     4  2
  _____
=  _____
```

**Prob.** 6: 13+25+35+12
Solution:

```
      1  3
+     2  5
+     3  5
+     1  2
  _____
=  _____
```

**Prob.** 7: 58+51+63+82
**Solution:**

```
      5  8
+     5  1
+     6  3
+     8  2
  _____

= _____
```

**Prob.** 8: 63+41+27+34
**Solution:**

```
      6  3
+     4  1
+     2  7
+     3  4
  _____

= _____
```

What have we done so far? We handled single digit numbers looking to get to 10. We learnt to decompose the digits into tens and excess or a deficiency. We handled the sum of tens separately and the small residual excess / deficiencies separately. We could do the entire operation mentally. With practice, we will attain speed and accuracy. This procedure also helped us to avoid the carry-over process explicitly. The process of looking at a set of numbers and understanding in what order we would like to add or manipulate determines the speed and accuracy.

The other common area we make mistakes in our regular class work, home assignments and school related academic material was when we were dealing with carry-overs. How did we want to deal with carry-overs? We wanted to have a mechanism of retaining carry-overs in every place until we finish the entire process of addition. While dealing with a large set of numbers, we ended up with a need for a notation to capture the carry-over digits as subscripts. We retained the carry-over right there and finally we resolved the carry-overs to create the final answer. In doing so we realized that the error in the carry-over process was avoided, significantly increasing the accuracy of the results.

## 2.4 Positive and Negative Numbers

Most folks have trouble dealing with negative numbers. From the first day at school, the concept of a positive number is intimately associated with the count of a physical object. For example, we refer to 3 apples, 6 chocolates and 2 pens. Now, when we introduce the concept of a negative number, the struggle commences, because we use a different representation of numbers. We use number line and mark the negative numbers to the left of zero. The need for a different representation for positive and negative numbers makes things worse for children who are expected to deal with positive and negative numbers. Let us spend a moment and clarify the concept.

Let us imagine a number line that extends on both sides, how long is of no academic interest. Now, imagine 0 to be the center of the line. All positive numbers are marked to the right of 0 and negative numbers to the left of it. This is the classical picture of a number line.

For a moment, imagine negative to mean "opposite of". Therefore, −4 is simply the opposite of +4. This means −4 is exactly the same distance from 0 as +4, but on the opposite side of 0.

Now, +4 is simply referred to as 4. We drop the + sign in front of it and treat it as positive. Therefore ++4 is the same as +++4, which is the same as 4. We simply drop all the + signs in front of a number.

Now, what is −+4? We drop the + sign; and we are left with −4; which is 4 on the other side of 0. Therefore, we can conclude that −+number simply means −number. Similarly, +−number, means −number. This is because we drop the +sign as before.

Let us go to the next step and get a sense of what --4 means. The first − sign means that the number is on the other side of 0, which is negative. And the second − asks us to go to the other side of zero, which is positive.

Therefore, we can conclude the following.
- An even number of − (negative) signs in front of any number makes the number positive.
- An odd number of − signs make the number negative.
- The number of + signs has no impact on the sign of the number, they can be ignored.

## 2.4.1 Exercises for practice

**Prob. 1:** +++4 =
**Prob. 2:** +++-4 =
**Prob. 3:** - - - + + - - ++4 =
**Prob. 4:** - +- +- +- +- +- +- +- +- +- +4 =

When we come across a negative number, there is no reason for panic. Stay calm, imagine the number line and go through the manipulation using the basics, even if you arrive at an answer that does not make physical sense.

For example, if you had 3 chocolates, and I took away 4 chocolates from you; you would be left with 3 - 4 = –1 chocolates. Yes, a –1 chocolate does not make sense; but it is still the correct answer to the question.

## 2.5 Complements

The whole number system is made up of 10 symbols, 0 to 9. Each of these symbols represents a number from zero to nine. All the numbers repeat in specific order after number 100, 1000 and so on. These numbers, which have a 1 followed by one or more numbers is called a base number in the decimal number system. We speak in terms of bases and complements. So, what are complements? Every number is either below or above a base number. Complement is simply the deficiency of a number from its base. A number 87 has a base of 100 and is 13 away from 100. Therefore, 13 is the complement of 100. Or, we can write the same thing as:

- Complement = Base – Number
- Number = Base – Complement
- Base = Complement + Number

Let us look at a few examples to get familiar with the concept of a base and complement of a number.

- Complement of 93 is 07; base is 100
- Complement of 65 is 35; base is 100
- Complement of 87 is 13; base is 100

One way of arriving at a base is to simply write 1 followed by as many zeros as the digits in the number.

- 87 is a two digit number, therefore the base number is 100 (one followed by two zeros)
- 93 is a two digit number, therefore the base number is 100 (one followed by two zeros)
- 987 is a three digit number, therefore the base number is 1000 (one followed by three zeros); and the complement of 987 is 013 which is simply the result of base minus the number.

### 2.5.1 Determining the complements of large numbers

The thumb rule is "all from nine and last from 10". This means, we subtract all the numbers from 9. The number in the Units place (the last number) is subtracted from 10.

Complement of 743 = (9–7)(9–4)(10–3) = 257
Complement of 8537 = (9–8)(9–5)(9–3)(10–7) = 1463
Complement of 19789 = (9–1)(9–9)(9–7)(9–8)(10–9) = 80211
Complement of 65319 = (9–6)(9–5)(9–3)(9–1)(10–9) = 34681

### 2.5.2 Summary of Rules for determining the complements

1. For finding the complement of any number we subtract individual digits.
2. Last digit has to be subtracted from 10 and all the others from 9.
3. If the number ends in 0 then
4. We just write 0 as the last digit of the complement also
5. We take the next digit from the right as the last digit and subtract it from 10.
6. All the other digits have to be subtracted from 9.

## 2.6 Vinculum Representation

Vinculum comes from a Latin word, it basically means to chain or to bond. We will see what we mean by the vinculum and how we plan to use this concept to ease numerical calculations.

Let us look at one common source of errors in our calculations. It happens when we handle large digits in numbers. A multiplication operation gives rise to carry forwards and things tend to get messy if you made a mistake in handling these things. If we have to reduce the probability of errors while handling large numbers, we need a method to address this issue. This is where the concept of vinculum comes in.

Let us take a look at the number 29. This is simply the sum of 2 tens and 9 units. We could also write the same number as 3 tens and –1 units.

---

Mathematically speaking:  $29 = 2 \times 10 + 9 \times 1 = 3 \times 10 + (-1) \times 1$

If we could put a –1 in the Units place and 3 in the Tens place, we would have a new representation for 29, which would not have the large digit 9 in it!

We use a small bar over 1 to indicate the fact that it is –1. And put this number in the units place. This number with a bar over it is "vinculum 1" which stands for –1. Therefore, –1 is represented as $\bar{1}$ and read as vinculum 1.

Hence:  $29 = 3\bar{1}$

How does this help us? Let consider a problem in multiplication.

**Ex. 3:** $29 \times 4$

Using conventional multiplication techniques,
1. *Step 1:* $9 \times 4 = 36$; Write down 6 and Carry 3 forward
2. *Step 2:* $4 \times 2 = 8$. $8 + 3$ (carry forward) = 11. Write down 11.
3. *Step 3:* Product is 116

Using vinculums:  $29 \times 4 = 3\bar{1} \times 4 = 12\bar{4} = 116$

Clearly, we do not have to handle the carry forwards. The complexity is reduced dramatically.

Let us look at a few examples of vinculum representation where we have tried to eliminate large digits in the numbers.

$$19 = 2\bar{1}$$
$$86 = 9\bar{4}$$
$$67 = 7\bar{3}$$
$$48 = 5\bar{2}$$

Let us now take a baby step forward. We will try to understand two important techniques in vinculum.
- Converting a regular number into a vinculum

- Converting a vinculum into a regular number

These two procedures will equip us with the necessary tools for handling vinculums in order to reduce the problem complexity.

## 2.6.1 Converting a regular number to a vinculum

We will look at few numbers and identify the underlying patterns

| # | Number | | Is nothing but | | Vinculum |
|---|--------|---|----------------|---|----------|
| 1 | 7 | = | 10 − 3 | = | $1\bar{3}$ |
| 2 | 8 | = | 10 − 2 | = | $1\bar{2}$ |
| 3 | 19 | = | 20 − 1 | = | $2\bar{1}$ |
| 4 | 36 | = | 40 − 4 | = | $4\bar{4}$ |
| 5 | 48 | = | 50 − 2 | = | $5\bar{2}$ |

We get the drift of what is really going on. We take a number and represent it as a deficiency.

For example:

| 48 | is | 2 | short of | 50; | vinculum is | $5\bar{2}$ |
|---|---|---|---|---|---|---|
| 27 | is | 3 | short of | 30; | vinculum is | $3\bar{3}$ |
| 26 | is | 4 | short of | 30; | vinculum is | $3\bar{4}$ |
| 47 | is | 3 | short of | 50; | vinculum is | $5\bar{3}$ |
| 58 | is | 2 | short of | 60; | vinculum is | $6\bar{2}$ |
| 22 | is | 8 | short of | 30; | vinculum is | $3\bar{8}$ |

So we get the general idea of how to get to a vinculum representation.

***Rule: Take the 10s complement of the number you want to vinculate; and increase the digit to its left by 1.***

Let us look at 27. Seven is a large digit; let us convert this into a vinculum. 10s complement of 7 is 3. So, we write this down as $\overline{3}$. We then increase the value of digit to its left by 1. 2+1=3. We write down $3\overline{3}$.

## 2.6.2 Converting a vinculum to a regular number

So, now we will do the inverse process and see how that works. In the previous set of examples, we said, 17 was 10 + 7, the same as 20 minus 3 therefore it is $2\overline{3}$. And therefore $2\overline{3}$ is the same as 20 minus 3, which is 17. Similarly, $1\overline{3}$ is 10 minus 3, which is 7. You can try out a few more examples to convince yourself. Therefore we can create the following rule.

***Rule: Take the 10s complement of the number under vinculum; and decrease the digit to its left by 1.***

Consider the number $2\overline{3}$. 10s complement of number under vinculum (which is 3) is 7. This is the units' place of the result. Now, we decrease the digit to its left by 1. This gives us 2 – 1 = 1. Therefore $2\overline{3}$ = 17.

Let us quickly go through a few examples to highlight the process and help you gain some valuable practice as well.

$\overline{5}8$ : 10s complement of 8 is 2; reduce 5 by 1; result is 42
$8\overline{3}$ : 10s complement of 3 is 7; reduce 8 by 1; result is 77
$\overline{4}9$ : 10s complement of 9 is 1; reduce 4 by 1; result is 31

Let us reinforce an observation that we made in this chapter. The reason why we typically want to vinculate is we want to deal with small single digit numbers. This means we typically want to vinculate numbers that are larger than 5. When we vinculate numbers like 6, 7, 8, 9 we get 4, 3, 2 and 1 respectively. These are small single digit numbers. On either side of 5, you will be left with 1, 2, 3, 4, 5, 4, 3, 2 and 1. We see that the numbers start becoming small and our ability to manipulate is vastly enhanced. So that is something that is worth a thought.

Mathematics is not quite like some of these wildlife documentary channels where you can sit on a couch, listen and absorb and get a general drift.

You become good at math by practice. It is almost like what perhaps someone like Roger Federer does with his tennis. It takes hours and hours of practice to reach a certain level of perfection. Speed Math requires that sort of dedication as well.

## 2.7   Digital Roots

Let us spend some quality time on the issue of Digital Roots. A typical academic challenge that we face is as follows. We follow a prescribed syllabus at school and learn our lessons. We then appear for a test. And we do make a few mistakes – silly mistakes as they are referred to. The response to an error in Math is very stereotypical to say the least. You are expected to solve the same problem all over again; chances are that you will make the same mistake one more time! There is a reason why this pattern recurs. It is because we are taught techniques to solve problems. Techniques for identifying errors are seldom taught. This topic is very rarely the focus area of any Math curriculum.

There is a way to check if your answer is OK. For making this important check happen, we need to understand the concept of digital roots. We will not define the concept of digital roots. Instead let us experience what this is by solving a few problems.

Let us consider the number 5672. Let us add all the digits together. Therefore, $5+6+7+2=20$. We arrive at a two-digit answer. Let us add the digits again, and continue to repeat this process until we come to a single digit sum. In other words, we will add the digits of 20: $2+0=2$. Therefore, the digital root of 5672 is 2. Some books refer to this concept as a digit sum. Whatever be the name, the underlying principle is the same. We start off with a number whose digital root is of interest. We then determine the sum of the digits of the number. If the sum is not a single digit number, we continue to add the digits of the sum together. The resultant single digit answer to this repeated addition of the digits of the sum is called the digital root of the number.

If this sounds confusing, I suggest we keep the faith and solve a few problems. The actual problem solving process is not that complicated. The explanation in English is! ☺

**Ex.** 1: Find the digital root of 1332.
Sum of digits of 1332: 1+3+3+2 = 9. This is a single digit number. So, the digital root is 9.

**Ex.** 2: Find the digital root of 2135.
Sum of digits of 2135: 2+1+3+5 = 11; 11 is a two digit number
Sum of digits of 11: 1+1 = 2. This is single digit number; so the digital root is 2.

**Ex.** 3: Find the digital root of 1858.
Sum of the digits of 1858: 1+8+5+8 = 22; 22 is a two digit number
Sum of digits of 22: 2+2 = 4. This is a single digit number; so the digital root is 4.

In other words, you take a number and repeatedly add the digits in the number or its digit sum, till you are left with a single digit number. That single digit number is referred to as the digital root of the number.

Now that we are familiar with what a digital root really stands for; we can now move on and look for ways to get to this faster. There are several ways of doing this.

Let us start looking at a few patterns. Digital root of 1, 2, 3, 4, 5, 6, 7, 8 and 9 is 1, 2, 3, 4, 5, 6, 7, 8 and 9 respectively. Digital root of 10 is 1, 11 is 2, 12 is 3 and so on. Therefore, we can represent this fact in a couple of ways. Let us look at the table below:

| Digital Roots | 1 | 2 | 3 | 4 | 5 | 6 | 7 | 8 | 9 |
|---|---|---|---|---|---|---|---|---|---|
| Numbers with | 1 | 2 | 3 | 4 | 5 | 6 | 7 | 8 | 9 |
| the same DR in | 10 | 11 | 12 | 13 | 14 | 15 | 16 | 17 | 18 |
| the same column | 19 | 20 | 21 | 22 | 23 | 24 | 25 | 26 | 27 |

Clearly numbers 1, 10, 19 and 28 have the same digital root. Similarly 2, 11, 20 and 29 have the same digital root. Now digital root of 18, 27, 36, 45, 54, 63, 72 and 81 is 9 or 0.

This really means while computing the digital root of a number, say, 634 – we can discard 6 and 3 which add up to a 9 and conclude that the digital root is 4.

There is another way of looking at this pattern. Divide the number by 9. 634 divided by 9 leaves a reminder of 4, which is the digital root.

Clearly, casting away the individual digits that add up to 9 will ease the process of computation. Let us use this technique and determine the digital root of 812763. We simply discard (8, 1), (7, 2) and (6, 3); since they all add up to 9. The digit sum is 0. When digit sum is zero, the digital root is 9. Therefore the digital root is 9.

In the number 982176, we can discard (9), (8, 1) and (7, 2); and we are left with 6. Therefore, the digital root is 6.

Let us consider 999978. We can discard all the 9s in the number. We are left with 87; therefore the digital root is 6.

## 2.7.1 Applications of Digital Roots

### 2.7.1.1 Addition:

Let us look at a simple problem in addition:    15 + 97 = 112

Let us now determine the digital roots of the numbers involved in this equation and identify a pattern.
- Digital Root of 15: 6
- Digital Root of 97: 7
- Digital Root of 112: 4

If you look closely, the sum of digital roots of 15 and 97 is equal to digital root of 112.
- Sum of Digital Roots of 15 and 97: 6 + 7 = 13.
- Digital Root of 13 = 4
- Digital Root of 112 = 4

We can see that *the digital root of sum of digital roots of numbers is equal to the digital root of sum of numbers.*

Conversely, *if this equality is established, we know that we have checked for accuracy of addition.*

We can use digital roots to check for accuracy of addition. Let us assume that we have made a mistake in the addition process and we have somehow come to the solution that $15 + 97 = 113$.

- The digital root of sum of digital roots of 15 and 97: 4
- The digital root of sum of 113 (sum of numbers): 5

This inequality suggests that we have made a mistake in addition and that we need to re-do the problem. This is how we use a completely different technique to check the accuracy of the solutions to problems.

## 2.7.1.2 Subtraction:

Let us consider an example in subtraction: $98 - 63 = 35$
Just like we did in the case of addition, let us look at the digital roots of the numbers in the left hand side and right hand side of the equation.

- Digital Root of $98 = 8$
- Digital Root of $63 = 0$
- Digital Root of $98$ – Digital Root of $63 = 8 - 0 = 8$
- Digital Root of $35 = 8$

We can see that *the digital root of difference of digital roots of numbers is equal to the digital root of difference of numbers.*

Conversely, *if this equality is established, we know that we have checked for accuracy of subtraction.*

If for some reason, you arrive at an inequality, we immediately know that we have made a mistake in subtraction.

## 2.7.1.3 Multiplication:

Let us now look at a problem in multiplication: $98 \times 97 = 9506$

Let us now determine the digital roots of the number on the left hand side. And find the digital root of the product of digital roots.

- The digital root of $98 = 8$
- The digital root of $97 = 7$
- The product of digital roots of 98 and 97 = 56

- The digital root of product of digital roots = 2
- The digital root of 9506 = 2

We can see that *the digital root of product of digital roots of numbers is equal to the digital root of product of numbers.*

Conversely, *if this equality is established, we know that we have checked for accuracy of multiplication.*

If for some reason, you arrive at an inequality, we immediately know that we have made a mistake in multiplication.

No matter how much you practice, you will see that you will make a mistake due to fatigue or momentary lapse in concentration or some other reason. It is important for us to start using the power of digital roots in order to identify the errors that creep into our solutions and take appropriate corrective action.

Let us summarize our findings in three easy tongue twisters.
1. The digital root of sum of digital roots of numbers is equal to the digital root of sum of numbers
2. The digital root of sum of digital roots of numbers is equal to the digital root of sum of numbers
3. The digital root of sum of digital roots of numbers is equal to the digital root of sum of numbers

## 2.7.2 Exercises for practice

Determine the digital roots by casting out 9s.

**Prob. 1:** 9781328
**Prob. 2:** 998799
**Prob. 3:** 812367
**Prob. 4:** 673524
**Prob. 5:** 897643
**Prob. 6:** 705936
**Prob. 7:** 732014
**Prob. 8:** 272
**Prob. 9:** 32636
**Prob. 10:** 422736

## 2.7.3 Exercises for practice

Number puzzles on digital roots.

| If the digital roots of a set of numbers between 1 and 99 is: | And if the numbers satisfy the condition | The numbers are: |
|---|---|---|
| 5 | Difference between the figures is 3 | |
| 6 | The figures are the same | |
| 6 | First figure is double the second | |
| 7 | Difference between the figures is 3 | |
| 7 | One figure is 4 | |
| 6 | Both figures are odd | |
| 5 | The figures are consecutive | |
| 9 | The figures are consecutive | |
| 3 | One figure is double the other | |
| 8 | The answer is below 20 | |
| 1 | The number is less than 40 | |
| 1 | The first figure is 2 | |

## 2.7.3.1 Divisibility

While dealing with the number line, we note that 1 is a basic number. This is in the sense that all other numbers can be generated or "created" by adding or subtracting several ones to and from it.

For example,

| | |
|---|---|
| $3 = 1+1+1$ | Repeated addition of 1 |
| $4 = 1+1+1+1$ | Repeated addition of 1 |
| $-5 = -1-1-1-1-1$ | Repeated subtraction of 1 |

Let us start with the single digit positive numbers. The numbers 2, 4, 6 and 8 are called even numbers. They are all divisible by 2. The numbers 1, 3, 5, 7 and 9 are not divisible by 2, therefore they are called odd numbers. We can now look at an extended pattern. All numbers that end in an even number are even; all numbers that do not end in an even number are odd. Numbers ending in 0 are even numbers. This basic pattern defines the test for divisibility of a number by 2. Is it odd or even? If it is even, then the number is divisible by 2; otherwise it is not divisible by 2.

A student of middle school algebra student will represent an even number by $2n$ saying that it is a multiple of 2 or it is divisible by 2. If your teacher asks you to write down a general form of even number, we write this as $2n$. A general form of an odd number is $2n-1$ or $2n+1$. Every even number on the number line is surrounded by 2 odd numbers and every odd number is surrounded by even numbers. Even and odd numbers alternate on the number line.

The properties of division by numbers are known as divisibility. The key question that the topic of divisibility asks is – why do we say a number is divisible by another number?

All numbers are divisible by 1. This is the identity property of division. Any number divided by 1 is the number itself.

Every even number is divisible by 2. Therefore, we can conclude that every number that ends in 0, 2, 4, 6 or 8 is an even number. We can also say that every number that ends in an even digit is an even number.

If the last two digits (numbers in the Tens and Units place) are divisible by 4, we can conclude that the number is divisible by 4. Why is this true? Let us consider a 3 digit number $xyz$. This number is the $100x + 10y + z$. $100x$ is divisible by 4; Therefore, $10y + z$ needs to be divisible by 4 for the number $xyz$ to be divisible by 4. 116 is divisible by 4 because 16 is divisible by 4. 992 is divisible by 4 because 92 is divisible by 4.

Similarly, we can extend the logic to divisibility by 8. If the last three digits of a number is divisible by 8, then the number is divisible by 8. Using the clue of how we approached to prove the divisibility by 4, we can extend the same logic to prove this statement as well. 1064 is divisible by 8 because 064 is divisible by 8. 1800 is divisible by 8 because 800 is divisible by

8. Is 9128 divisible by 8? 128, the number formed by the last three digits, is divisible by 8. Therefore 9128 is divisible by 8.

If a number ends in 0 or 5, then the number is divisible by 5. 50 is divisible by 5 because it ends in a 0. 135 is divisible by 5 because it ends in 5. If the last two digits of a number is divisible by 25, then the number is divisible by 25 and similarly, if the last three digits of a number is divisible by 125, we can conclude that the number is divisible by 125. The pattern is similar to divisibility by 4 and 16.

That brings us to divisibility by 3. Three has a unique test for divisibility. If the num of the digits is divisible by 3, then the number is divisible by 3. For example, 216 is divisible by 3. This is because the sum of the digits of 216 is 9. And 9 is divisible by 3. 411 is divisible by 3, because the sum of the digits of the number is 6. And 6 is divisible by 3.

Now we will cover divisibility by 3-squared or 9. The test is similar to divisibility by 3. If the sum of the digits of the number is divisible by 9, then the number is divisible by 9. 729 is divisible by 9. The sum of the digits of the number is $7+2+9=18$. 18 is divisible by 9; therefore we can conclude that 729 is divisible by 9.

The divisibility by 6, throws up a very interesting property. If a number is divisible by 3 and 2, then the number is divisible by 6. In other words, if 6 is a factor of a number, this automatically means that 2 and 3 are their factors as well. This means that the number must pass the test of divisibility by 2 and 3. Similarly, if a number is divisible by 3 and 4, then the number is divisible by 12.

We have left 7 out of the entire discussion. It is beyond the scope of the discussion. The idea is to look at divisibility as a set of useful number patterns that emerge and have a sense of what this pattern really stands for.

In summary, we can make a few statements with a view to ease the pain of comprehension.
1.  1 is not divisible.
2.  Every number that is not a prime number is divisible.
3.  Prime number is divisible by itself and 1. Purely from a divisibility standpoint, we will consider them as indivisible.
4.  Even numbers end with 2, 4, 6, 8 and 0

5. Odd numbers end with 1, 3, 5, 7 and 0

The tests of divisibility have been captured in the table below. It should serve as a ready reference.

| Number | Property of Divisibility |
|---|---|
| 1 | All Numbers are divisible by 1. Period. No ifs, ands or buts. |
| 2 | Numbers divisible by 2 are even numbers, therefore end in 2, 4, 6, 8 or 0 |
| 3 | If the sum of the digits in a number is divisible by 3, the number is divisible by 3 |
| 4 | If the last two digits of a number is divisible by 4, then the number is divisible by 4 |
| 5 | Numbers divisible by 5 end with 0 or 5 |
| 6 | Numbers divisible by 3 and 2 are divisible by 6 |
| 7 | Let us skip this for now. |
| 8 | If the last three digits of a number is divisible by 8, then the number is divisible by 8 |
| 9 | If the sum of the digits in a number is divisible by 9, the number is divisible by 9 |
| 10 | Numbers divisible by 10 end with 0 |
| 100 | Numbers ending with 00 are divisible by 100 |
| 25 | If last two digits of a number is divisible by 25, then the number is divisible by 25 |

# 2.8   Representation of Numbers

One common source of mistakes during our problem solving process happens when we handle borrows and carry forwards. We need to devise a mechanism for handling these situations in a trouble-free way. We will simply extend the decimal number system that we are so familiar with.

In the representation of numbers in the decimal number system, every time the place value has a number greater than 9, we leave the units digit in the current place value and carry forward the other digits to the next place. This statement in English can be experienced as follows.

**Ex. 1:** 169 + 2
**Solution:**

| H | T | U | Place Values |
|---|---|---|---|
|   | 1 |   | Carry Forward |
| 1 | 6 | 9 | First Number |
| + |   | 2 | Second Number |
| = 1 | 7 | 1 | Sum |

We start with the Units place. 9+2 = 11. This is greater than 9; so we leave 1 in the units place and carry over 1 to the 10s place.

In the Tens place, we add the carry forward number, 1 to 6 and we get 7. Then we bring down the Hundreds digit as is, since there is no need to add a number here.

The answer is 171.
Instead, we will use a two-pass approach to this problem. We will solve it as follows. We will not use a carry forward like the way we did before.

| H | T | U | Place Values |
|---|---|---|---|
| 1 | 6 | 9 | First Number |
| + |   | 2 | Second Number |
| 1 | 6 | $_1$1 | Sum with subscripted carryovers |
| = 1 | 7 | 1 | Sum with collapsed carryovers |

This basic adjustment in representation of numbers will enable us to complete the computation quickly and adjust for carry forwards in one go.

Let us consider another example. Now, we will look at an example in multiplication.

| Th | H | T | U |
|---|---|---|---|
|   | 2 | 3 | 9 |
| × |   |   | 4 |
|   | 8 | $_1$2 | $_3$6 |
| = |   | 9 | 5 | 6 |

Doing this in the second pass enables us to write the answer from left to right. We can complete the multiplication from left to right as well, because we are only bothered about the product and not about the resultant carry forward at every intermediate stage. This helps us to gain speed. The accuracy component comes from 'taking a deep breath and handling the mess of carry forwards in one go''.

## 2.8.1 Summary

This brings us to the end of the chapter that deals with all the introductory concepts. It's a good idea to summarize our key highlights and observations. As we navigated through positive and negative numbers, divisibility, complements, vinculums and digital roots, our core motivation remained the same.

How can we make numerical manipulations simpler?

How can we reduce our probability of errors and silly mistakes?

As we solved problems, we saw a few interesting patterns emerge. These patterns lead us to the formulation of key techniques and rules. The patterns also helped us get familiar with the language of numbers.

We always looked at ways of dealing with small numbers. We took every opportunity to identify ways and means to eliminate complicated operations such as carry forward, borrow and others.

We wanted to explore ways to check our solution space for errors. We found a method for using digital roots as a way of accomplishing this. They say to err is human and to detect is to use digital roots.
We saw the elegance of alternative number representation to decimal representation. We used vinculum representation. We also saw how we could go back and forth and manipulate numbers.

Now we are equipped with a bunch of fundamentals that we can reuse as we go through other modules in our Speed Math journey.

# 3 Speed Addition

In the conventional addition technique, we write down the numbers we wish to add one below the other, with the numbers in each place value properly aligned. We count all the numbers in the Units place and write the sum of the digits in the Units column and adjust the Tens place for a carry-over digit. We proceed systematically from right to left; repeating the procedure we just discussed.

If we had to add a set of numbers with a decimal point, we align the decimal points; start from the right most column after the decimal point. We progressively move to the columns to its left and complete the addition process.

**Ex. 1:** 66.8 + 26.5 + 19.41 + 8.08
**Solution:** We go through the following steps.
*Step 1*: Write down the numbers one below the other, aligning the decimal points.

```
      6  6  .  8
+     2  6  .  5
+     1  9  .  4  1
+        8  .  0  8
=  _____
```

*Step 2*: We now add the numbers, just like we would add whole numbers beginning with the right-most column.

*Step 3*: We must remember to bring down the decimal point right below the decimal point in the problem. In other words, the decimal point of the sum is right below the decimal points of the numbers.

```
      6  6  .  8
+     2  6  .  5
+     1  9  .  4  1
+        8  .  0  8
=  1  2  0  .  7  9
```

Let us make a minor adjustment to the way we add, which will improve the speed and accuracy immensely. In order to understand this method, we will simply use an example.

**Ex. 2:** 8764 + 7654 + 3965 + 4321 + 6008
**Solution:**

| | Th | H | T | U |
|---|---|---|---|---|
| | 8 | 7 | 6 | 4 |
| + | 7 | 6 | 5 | 4 |
| + | 3 | 9 | 6 | 5 |
| + | 4 | 3 | 2 | 1 |
| + | 6 | 0 | 0 | 8 |
| = | | | | |

We will add three more rows and label them as:

**Running total:**
**Ticks:**
**Answer:**

We will use the rules:

1.  "Never count higher than eleven".
2.  When your sum cross eleven, add a small tick next to the digit and continue the counting process with the excess over eleven.
3.  The ticks represent the number of ticks in each column. In other words, this is simply the number of times our current total crossed eleven.
4.  The running total is the final sum that we are left with at the end of this process of not counting over eleven.

We will finally figure out how to use the ticks and running total information to complete our addition.

Let us consider an example to detangle this.

Let us add 6, 7, 5, 4 and 9 together.

|  |  |  |
|---|---|---|
|  | 6 |  |
|  | 7* |  |
|  | 5 |  |
|  | 4* |  |
|  | 9 |  |
| Ticks | 2 | [we have used a * to indicate a tick] |
| **Running total** | 9 |  |

We would start with 6 on the top. 6 plus 7 is 13. This is greater than 11; so we subtract 11 from 13; this gives us 2 as our current running total and we put a tick next to 7. Now we add 2, 5 and 4 together. Now this gives us 11. Therefore our running total is 0 and we put a tick next to 4. That leaves 9 as the final running total. We have one tick next to 7 and another next to 4. Therefore we have 2 ticks in all.

Let us now return to Example 2 and see how we can make progress in our speed addition technique. We will write down the running totals and ticks based on the short example we just looked at.

|  |  | Th | H | T | U |
|---|---|---|---|---|---|
|  |  | 8 | 7 | 6 | 4 |
| + |  | 7* | 6* | 5* | 4 |
| + |  | 3 | 9* | 6 | 5* |
| + |  | 4* | 3 | 2 | 1 |
| + |  | 6 | 0 | 0 | 8* |
| Running Total | 0 | 6 | 3 | 8 | 0 |
| Ticks | 0 | 2 | 2 | 1 | 2 |

Now we arrive at the final result by add a zero in front of the row of running total and ticks. We add the neighbor on the right in the bottom row of ticks to the sum of ticks and running total in a particular column. For example:

|  |  |  |  |  |  |
|---|---|---|---|---|---|
| − | − | − | 8 | − |  |
| − | − | − | 1 | 2 |  |
| − | − | − | 11 | − |  |

Here: 8+1+2=11

Considering the next column, we would get:

$$- \quad - \quad 3 \quad - \quad -$$
$$- \quad - \quad 2 \quad 1 \quad -$$
$$- \quad - \quad 6 \quad - \quad -$$

Here: 3+2+1=6

Thus, we can now consider the entire row of running totals and ticks as below.

| Running Total | 0 | 6 | 3 | 8 | 0 |
|---|---|---|---|---|---|
| Ticks | 0 | 2 | 2 | 1 | 2 |
| Answer | | 2 | ₁0 | 6 | ₁1 | 2 |
| = | | 3 | 0 | 7 | 1 | 2 |

Let us now summarize the key aspects of the technique.
1. Elevens Rule: Never count higher than eleven - When a total crosses eleven, we subtract 11 from the running total and add a tick.
2. Subtract eleven means, drop the tens digit and reduce the units digit by 1. This is the pattern that we notice.
3. Put down the running totals and ticks in two rows.
4. Add a 0 in front of both the rows of running totals and ticks.
5. Calculate in L. "Running Total + Ticks in that Column + Ticks to the right of it". This is the final answer.

**Ex.** 3: 7865 + 6675 + 5463 + 45 + 765
**Solution:**

|  |  | 7 | 8 | 6 | 5 |  |
|---|---|---|---|---|---|---|
| + |  | 6 * | 6 * | 7 * | 5 |  |
| + |  | 5 | 4 | 6 | 3 * |  |
| + |  |  |  | 4 * | 5 |  |
| + |  |  | 7 * | 6 | 5 * |  |
| Running Total | 0 | 7 | 3 | 7 | 1 |  |
| Ticks | 0 | 1 | 2 | 2 | 2 |  |
|  |  |  |  |  | 3 | Step 1: 1+2 |
|  |  |  |  | $_1$1 |  | Step 2: 7+2+2 |
|  |  |  | 7 |  |  | Step 3: 3+2+2 |
|  |  | $_1$0 |  |  |  | Step 4: 7+1+2 |
|  | 1 |  |  |  |  | Step 5: 0+0+1 |
|  | 1 | $_1$0 | 7 | $_1$1 | 3 |  |
| = | 2 | 0 | 8 | 1 | 3 |  |

**Ex.** 4: 876 + 6754 + 650 + 24 + 376
**Solution:**

|   |   |   |   | 8 | 7 | 6 |   |
|---|---|---|---|---|---|---|---|
| + |   |   | 6 | 7* | 5* | 4 |   |
| + |   |   |   | 6 | 5 | 0 |   |
| + |   |   |   |   | 2 | 4* |   |
| + |   |   |   | 3* | 7* | 6 |   |
| Running Total |   | 0 | 6 | 2 | 4 | 9 |   |
| Ticks |   | 0 | 0 | 2 | 2 | 1 |   |
|   |   |   |   |   |   | $_1$0 | Step 1: 9+1 |
|   |   |   |   |   | 7 |   | Step 2: 4+2+1 |
|   |   |   |   | 6 |   |   | Step 3: 2+2+2 |
|   |   |   | 8 |   |   |   | Step 4: 6+0+2 |
|   |   |   | 8 | 6 | 7 | $_1$0 |   |
| = |   |   | 8 | 6 | 8 | 0 |   |

**Ex. 5:** 78.9 + 321 + 45.8 + 760 + 59.9
**Solution:**

|   |   |   |   | 7 | 8 | . | 9 |
|---|---|---|---|---|---|---|---|
| + |   |   | 3 | 2 | 1 | . | 0 |
| + |   |   |   | 4* | 5* | . | 8* |
| + |   |   | 7 | 6 | 0 | . | 0 |
| + |   |   |   | 5* | 9* | . | 9* |
| Running Total |   | 0 | $_1$0 | 2 | 1 | . | 4 |
| Ticks |   | 0 | 0 | 2 | 2 | . | 2 |
|   |   | 0 | $_1$2 | 6 | 5 | . | 6 |
| = |   | 1 | 2 | 6 | 5 | . | 6 |

**Ex. 6:** 987 + 765 + 64 + 77 + 8 + 565 + 432 + 876
**Solution:**

| | | | | |
|---|---|---|---|---|
| | | 9 | 8 | 7 |
| | | 7* | 6* | 5* |
| + | | | 6 | 4 |
| + | | | 7* | 7* |
| + | | | | 8 |
| + | | 5 | 6* | 5* |
| + | | 4* | 3 | 2 |
| + | | 8* | 7 | 6* |
| Running Total | 0 | 0 | $_{1}0$ | 0 |
| Ticks | 0 | 3 | 3 | 4 |
| | 3 | 6 | $_{1}7$ | 4 |
| = | | 3 | 7 | 7 | 4 |

**Ex. 7:** 78654 + 70543 + 65489 + 7777 + 888 + 99
**Solution:**

| | | | | | | |
|---|---|---|---|---|---|---|
| | | 7 | 8 | 6 | 5 | 4 |
| + | | 7* | 0 | 5* | 4 | 3 |
| + | | 6 | 5* | 4 | 8* | 9* |
| + | | | 7 | 7* | 7* | 7* |
| + | | | | 8 | 8 | 8 |
| + | | | | | 9* | 9* |
| Running Total | 0 | 9 | 9 | 8 | 8 | 7 |
| Ticks | 0 | 1 | 1 | 2 | 3 | 3 |
| | 1 | $_{1}1$ | $_{1}2$ | $_{1}3$ | $_{1}4$ | $_{1}0$ |
| = | 2 | 2 | 3 | 4 | 5 | 0 |

**Ex. 8:** 9878 + 675 + 23.6 + 832.9 + 22 + 721.6 + 67
**Solution:**

| | | | | | | . | |
|---|---|---|---|---|---|---|---|
| | | 9 | 8 | 7 | 8 | . | 0 |
| + | | | 6* | 7* | 5* | . | 0 |
| + | | | | 2 | 3 | . | 6 |
| + | | | 8* | 3 | 2 | . | 9* |
| + | | | | 2 | 2 | . | 0 |
| | | | 7 | 2* | 1 | . | 6 |
| + | | | | 6 | 7* | . | 0 |
| Running Total | 0 | 9 | 7 | 7 | 6 | . | $_{1}0$ |
| Ticks | 0 | 0 | 2 | 2 | 2 | . | 1 |
| | 0 | $_{1}1$ | $_{1}1$ | $_{1}1$ | 9 | . | $_{1}1$ |
| = | 1 | 2 | 2 | 2 | 0 | . | 1 |

**Ex. 9:** 6785 + 5643 + 5432 + 7676 + 8998
**Solution:**

| | | | | | |
|---|---|---|---|---|---|
| | | 6 | 7 | 8 | 5 |
| + | | 5* | 6* | 4* | 3 |
| + | | 5 | 4 | 3 | 2 |
| + | | 7* | 6* | 7* | 6* |
| + | | 8 | 9 | 9 | 8* |
| Running Total | 0 | 9 | $_{1}0$ | 9 | 2 |
| Ticks | 0 | 2 | 2 | 2 | 2 |
| | 2 | $_{1}3$ | $_{1}4$ | $_{1}3$ | 4 |
| = | 3 | 4 | 5 | 3 | 4 |

## 3.1.1 Exercises for practice

**Prob. 1:** 0.987 + 1.578 + 2.09 + 56.78 + 0.675 + 32.9 + 77
**Solution:**

|   |   |   | 0 | . | 9 | 8 | 7 |
|---|---|---|---|---|---|---|---|
| + |   |   | 1 | . | 5 | 7 | 8 |
| + |   |   | 2 | . | 0 | 9 |   |
| + |   | 5 | 6 | . | 7 | 8 |   |
| + |   |   | 0 | . | 6 | 7 | 5 |
|   |   | 3 | 2 | . | 9 |   |   |
|   |   | 7 | 7 | . | 0 |   |   |

Running Total

Ticks

=

**Prob. 2:** 897 + 543 + 765 + 342 + 670
**Solution:**

|   | 8 | 9 | 7 |
|---|---|---|---|
| + | 5 | 4 | 3 |
| + | 7 | 6 | 5 |
| + | 3 | 3 | 2 |
| + | 6 | 7 | 0 |

Running Total

Ticks

=

Basics of Speed Mathematics

**Prob. 3:** 1234 + 6004 + 786 + 65 + 89076 + 564 + 54321 + 88
**Solution:**

|   |   |   | 1 | 2 | 3 | 4 |
|---|---|---|---|---|---|---|
| + |   |   | 6 | 0 | 0 | 4 |
| + |   |   |   | 7 | 8 | 6 |
| + |   |   |   |   | 6 | 5 |
| + |   | 8 | 9 | 0 | 7 | 6 |
| + |   |   |   | 5 | 6 | 4 |
| + |   | 5 | 4 | 3 | 2 | 1 |
| + |   |   |   |   | 8 | 8 |

Running Total

Ticks

=

**Prob. 4:** 787.787 + 654.321 + 333.222 + 564.876 + 876.333 + 822.321
**Solution:**

|   | 7 | 8 | 7 | . | 7 | 8 | 7 |
|---|---|---|---|---|---|---|---|
| + | 6 | 5 | 4 | . | 3 | 2 | 1 |
| + | 3 | 3 | 3 | . | 2 | 2 | 2 |
| + | 5 | 6 | 4 | . | 8 | 7 | 6 |
| + | 8 | 7 | 6 | . | 3 | 3 | 3 |
| + | 8 | 2 | 2 | . | 3 | 2 | 1 |

Running Total

Ticks

=

**Prob. 5:** 8765 + 4545 + 7543 + 6965 + 8899
**Solution:**

|   |   |   |   |   |
|---|---|---|---|---|
|   | 8 | 7 | 6 | 5 |
| + | 4 | 5 | 4 | 5 |
| + | 7 | 5 | 4 | 3 |
| + | 6 | 9 | 6 | 5 |
| + | 8 | 8 | 9 | 9 |

Running Total

Ticks

=

**Prob. 6:** 8867 + 55 + 32 + 136 + 9000 + 78 + 8932
**Solution:**

|   |   |   |   |   |
|---|---|---|---|---|
|   | 8 | 8 | 6 | 7 |
| + |   |   | 5 | 5 |
| + |   |   | 3 | 2 |
| + |   | 1 | 3 | 6 |
| + | 9 | 0 | 0 | 0 |
| + |   |   | 7 | 8 |
| + | 8 | 9 | 3 | 2 |

Running Total

Ticks

=

**Prob. 7:** 6754.87 + 809.76 + 564.9 + 23 + 881.8 + 76.5
**Solution:**

```
              6  7  5  4  .  8  7
    +            8  0  9  .  7  6
    +            5  6  4  .  9
    +               2  3  .  0
    +            8  8  1  .  8
    +               7  6  .  5
```

Running Total

Ticks

=

**Prob. 8:** 765 + 989 + 764 + 432 + 43 + 9
**Solution:**

```
              7  6  5
    +         9  8  9
    +         7  6  4
    +         4  3  2
    +            4  3
    +               9
```

Running Total

Ticks

=

## 3.2 The issue of accuracy

We will use every piece of information that we have so far. This will mean that we will use:

a. The column of figures
b. The rows of ticks and running totals; we will call these two rows as the working table.
c. The answer itself

For this we will determine:

a. The digital roots of the columns of figures
b. The digital roots for the working table
c. The digital root for the final answer

Let us consider a simple example to highlight the error checking process.

**Ex. 1:** 8764 + 7654 + 3965 + 4321 + 56 + 275 + 6008 + 6
**Solution:**

|  |  | 8 | 7 | 6 | 4 |
|---|---|---|---|---|---|
| + |  | 7 * | 6 * | 5 * | 4 |
| + |  | 3 | 9 * | 6 | 5 * |
| + |  | 4 * | 3 | 2 | 1 |
| + |  |  |  | 5 * | 6 |
| + |  |  | 2 | 7 | 5 * |
| + |  | 6 | 0 | 0 | 8 * |
| + |  |  |  |  | 6 |
| Running Total | 0 | 6 | 5 | 9 | 6 |
| Ticks | 0 | 2 | 2 | 2 | 3 |
|  | 2 | ₁0 | 9 | ₁4 | 9 |
| Column Roots |  | 1 | 0 | 4 | 3 |
| = | 3 | 1 | 0 | 4 | 9 |

Notice that we have underlined the figures that add up to a 9 within each column. This is a way of casting away the 9s while determining the digital roots. Such simple markings will help aid the problem solving process immensely.

Now we need to determine the check figure for the working table. For doing this, we simply add the digits in each column of the working table with twice the digits in the ticks.

| Running Total | 0 | 6 | 5 | 9 | 6 |
|---|---|---|---|---|---|
| Ticks | 0 | 2 | 2 | 2 | 3 |
| Ticks | 0 | 2 | 2 | 2 | 3 |
| Working Table | 0 | 1 | 0 | 4 | 3 |

Since the Column Roots and the Working Table rows match, we can now be sure that we are headed the right way. The column where we find a mismatch has the error in computation. We can quickly check our working in that column alone.

The final step of checking for accuracy is to compute the digital roots of check figure above and the final answer. So, this is what we have:

| Column Roots: | 01043 |
|---|---|
| Digital Root: | 8 |
| Answer: | 31049 |
| Digital Root: | 8 |

We see that the digital roots match. Therefore our addition is accurate.

**Summary of the error checking technique:**

1. Determine the digital roots for each of the columns. Use an underscore or some such marking for the set of numbers that add up to 9. This will help out cast away 9s easily; thus making the process of determining the digital roots easy.
2. Determine the check figure of the working table. This is simply the columnar sum of running total and twice the ticks in that column. Make a marking for the numbers that add up to 9 to make the process of determining the digital roots error free and easy.
    a. The two check figure rows must be identical if the addition is correct.

        b.   The column with the mismatch (if any) in the digital roots contains the error in addition. Fix it.

3.   The final step is simple

        a.   Find the digital root of the numbers in the check figure row

        b.   And the digital root of the final answer

4.   If these match, your addition is accurate.

Now, we are back to our number gym for additional practice.

**Ex. 2:** 876 + 6754 + 650 + 24 + 376
**Solution:**

|  |  |  |  | 8 | 7 | 6 |
|---|---|---|---|---|---|---|
| + |  |  | 6 | 7* | 5* | 4 |
| + |  |  |  | 6 | 5 | 0 |
| + |  |  |  |  | 2 | 4* |
| + |  |  |  | 3* | 7* | 6 |
| Running Total | 0 | 6 | 2 | 4 | 9 |  |
| Ticks | 0 | 0 | 2 | 2 | 1 |  |
|  |  |  | 8 | 6 | 7 | ₁0 |
| = |  |  | 8 | 6 | 8 | 0 |
| Column Roots |  | 6 | 6 | 8 | 2 |  |
| Working Table |  | 6 | 6 | 8 | 2 |  |

Column Roots and Working Table match. Good start ☺
Digital root of answer:        8+6+8+0 = 22; 2+2 = 4
Digital root of Working Table:   6+6+8+2 = 22; 2+2 = 4

Digital roots match. So the answer is accurate. ☺

**Ex. 3:** 78.9 + 321 + 45.8 + 760 + 59.9
**Solution:**

|                   |   |                  | 7 | 8 | . | 9  |
|-------------------|---|------------------|---|---|---|----|
| +                 |   |                  | 3 | 2 | 1 | . | 0 |
| +                 |   |                  |   | 4* | 5* | . | 8* |
| +                 |   |                  | 7 | 6 | 0 | . | 0 |
| +                 |   |                  |   | 5* | 9* | . | 9* |
| Running Total     | 0 | $_1$0            | 2 | 1 | . | 4  |
| Ticks             | 0 | 0                | 2 | 2 | . | 2  |
|                   | 0 | $_1$2            | 6 | 5 | . | 6  |
| =                 | 1 | 2                | 6 | 5 | . | 6  |
| Column Roots      |   | 1                | 6 | 5 |   | 8  |
| Working Table     |   | 1                | 6 | 5 |   | 8  |

Column Roots and Working Table match. We are almost there ☺

Digital Root for Working Table:     2
Digital Root for Answer:            2

The digital roots match. This is accurate indeed!!!

**Ex. 4:** 987 + 765 + 64 + 77 + 8 + 565 + 432 + 876
**Solution:**

|                  |   |   | 9 | 8 | 7 |
|------------------|---|---|---|---|---|
| +                |   |   | 7 * | 6 * | 5 * |
| +                |   |   |   | 6 | 4 |
| +                |   |   |   | 7 * | 7 * |
| +                |   |   |   |   | 8 |
| +                |   |   | 5 | 6 * | 5 * |
| +                |   |   | 4 * | 3 | 2 |
| +                |   |   | 8 * | 7 | 6 * |
| Running Total    | 0 | 0 | ₁0 | 0 |
| Ticks            | 0 | 3 | 3 | 4 |
|                  | 3 | 6 | ₁7 | 4 |
| =                | 3 | 7 | 7 | 4 |
| Column Roots     |   |   | 6 | 7 | 8 |
| Working Table    |   |   | 6 | 7 | 8 |

Digital root of Working Table:     3
Digital root of Answer:            3
Complete match ☺ Accuracy guaranteed!!!

**Ex. 5:** 78654 + 70543 + 65489 + 7777 + 888 + 99 + 8787
**Solution:**

|  |  | 7 | 8 | 6 | 5̲ | 4 |
|---|---|---|---|---|---|---|
| + |  | 7 * | 0 | 5̲ * | 4̲ | 3 |
| + |  | 6 | 5 * | 4̲ | 8 * | 9 * |
| + |  |  | 7 | 7 * | 7 * | 7 * |
| + |  |  |  | 8 | 8 | 8 |
| + |  |  |  |  | 9̲ * | 9̲ * |
| + |  |  | 8 * | 7 * | 8 * | 7 * |
| Running Total | 0 | 9 | 6 | 4 | 5 | 3 |
| Ticks | 0 | 1 | 2 | 3 | 4 | 4 |
|  | 1 | ₁2 | ₁1 | ₁1 | ₁3 | 7 |
| = | 2 | 3 | 2 | 2 | 3 | 7 |
| Column Roots |  | 2 | 1 | 1 | 4 | 2 |
| Working Table |  | 2 | 1 | 1 | 4 | 2 |

Digital root of Working Table:  1
Digital root of Answer:  1

The working is correct ☺

**Ex. 6:** 9878 + 675 + 23.6 + 832.9 + 22 + 721.6 + 67
**Solution:**

|  |  |  |  |  |  |  |  |
|---|---|---|---|---|---|---|---|
|  |  | $\underline{9}$ | 8 | $\underline{7}$ | $\underline{8}$ | . | 0 |
| + |  |  | 6* | 7* | 5* | . | 0 |
| + |  |  |  | $\underline{2}$ | 3 | . | 6 |
| + |  |  | 8* | $\underline{3}$ | 2 | . | $\underline{9}$* |
| + |  |  |  | $\underline{2}$ | $\underline{2}$ | . | 0 |
| + |  |  | 7 | 2* | $\underline{1}$ | . | 6 |
| + |  |  | $\underline{6}$ | $\underline{7}$* | . | 0 |  |
| Running Total | 0 | 9 | 7 | 7 | 6 | . | $_1 0$ |
| Ticks | 0 | 0 | 2 | 2 | 2 | . | 1 |
|  | 0 | $_1 1$ | $_1 1$ | $_1 1$ | 9 | . | $_1 1$ |
| = | 1 | 2 | 2 | 2 | 0 | . | 1 |
| Column Roots |  | 0 | 2 | 2 | 1 |  | 3 |
| Working Table |  | 0 | 2 | 2 | 1 |  | 3 |

| | |
|---|---|
| Digital root of Working Table: | 8 |
| Digital root of Answer: | 8 |

The answer is correct.

## 3.2.1   Exercises for practice

**Prob. 1:** 6785 + 5643 + 5432 + 7676
**Solution:**

```
              6  7  8  5
+             5  6  4  3
+             5  4  3  2
+             7  6  7  6
          ─────────────────
Running Total
Ticks


          ─────────────────
=
          ─────────────────
Column Roots
Working Table
```

**Prob. 2:** 0.987 + 1.578 + 2.09 + 56.78 + 0.675 + 32.9 + 77
**Solution:**

```
              0  .  9  8  7
              1  .  5  7  8
              2  .  0  9
+          5  6  .  7  8
+             0  .  6  7  5
+          3  2  .  9
+          7  7  .  0
          ──────────────────
Running Total
Ticks


          ──────────────────
=
          ──────────────────
Column Roots
Working Table
```

**Prob.** 3: 897 + 543 + 765 + 342 + 670 + 963 + 238
**Solution:**

```
                          8   9   7
              +           5   4   3
              +           7   6   5
              +           3   4   2
              +           6   7   0
              +           9   6   3
              +           2   3   8
              ─────────────────────
              Running Total
              Ticks


              ─────────────────────
              =
              ─────────────────────
              Column Roots
              Working Table
```

**Prob. 4:** 1234 + 6004 + 786 + 65 + 89076 + 564 + 54321 + 88
**Solution:**

```
                      1   2   3   4
              +       6   0   0   4
              +           7   8   6
              +               6   5
              +   8   9   0   7   6
              +           5   6   4
              +   5   4   3   2   1
              +               8   8
              ─────────────────────
              Running Total
              Ticks


              ─────────────────────
              =
              ─────────────────────
              Column Roots
              Working Table
```

## 3.3 Summary

We started with a simple thumb rule of not counting over 11; this means that we did not have to remember large numbers and spend anxious moments while adding numbers. By using the concept of a running total, we converted large numbers to small single digit numbers. Little ticks helped us to track the number of 11s in the columnar sums. Finally, we exploited the theory of digital roots to ensure accuracy. We found a way of determining the column in which an error would have crept up when we find a discrepancy in the digital root signatures that we called check figures. A bunch of cute little tricks – results in a giant leap in ease and accuracy.

# 4   Speed Subtraction

In this chapter, we will now turn our attention to subtraction. Although, we have used the word "speed" subtraction, we will focus on accuracy. Speed will simply be a by-product of the process.

Let us quickly recap the process of subtraction. In subtraction we are looking at the difference of two numbers. For example, 6 − 5 is 1, correct. 5 − 3 is 2 again. 9 − 6 is 3.

In other words, if the first number is larger than the second number, the subtraction process is straightforward. We typically get this right. In other words, problems such as 99 − 33 is 66; 9 − 3 is 6, 9 − 3 is 6 are simple. In this class of subtraction problems, probability of error is small. We can typically get it right with a little practice. There is no complication with that process.

The focus of the chapter is exclusively related to the other case - the second number is larger than the first number. In these cases, we are left with a negative number, which forces a borrow operation to ensure a positive output during the intermediate stages.

Let us consider, 14 − 6. You borrow one from the tens place. The units place now has 14 units in them. Now you subtract 14 − 6 to get 8. Since we borrowed 1 from the tens place, we are left with a zero in tens place. The process of subtraction comes to an end. This is how we traditionally solve the problem of subtraction.

The most probable place for errors in subtractions comes in two places:
   a.   We are subtracting a larger number from a smaller number; typically most of us have a discomfort in dealing with negative numbers.
   b.   The borrow operation leads to adding a ten to the current place value; and subtracting one from the adjacent place value. We are prone to making mistakes while we do this.

If we have to eliminate or reduce the probability of errors, we must make amends and get this process right.

Let us start with an example. We will start understanding the underlying patterns as we go through this process.

**Ex. 1:** 365 – 186

```
      3  6  5
  -   1  8  6
  =
```

The first step in the process is to make a small mark on each of the column of numbers where the number on the top is lesser than the number in the bottom. This will lead us to the following representation.

```
            *   *
      3  6  5
  -   1  8  6
  =
```

Let us now start the process of subtraction. Although you will be tempted to start off solving "*5 - 6 is less than zero so, let us borrow …*", I would us like to stop and resist the temptation to do so. Instead, we will do the following:

Rule 1: When the number on top is lesser than the number at the bottom, subtract the number at the top from the number at the bottom. In other words, "subtract larger number minus smaller number" and take the 10s complement of the result.

Therefore, we perform the following steps:
- 5 = 1
- 10s complement of 1 is 9 [10 – 1 = 9]

So, we write down 9 in the units place.

```
            *   *
      3  6  5
  -   1  8  6
  =            9
```

Now we come to the second column or the tens place. We see a marking on top of the column indicating that the first number is lesser than the second. We find that six is less than eight. We will repeat the steps as before, with a **minor** difference. We will subtract smaller number from the larger number and take 9s complement, instead of 10s complement. We take the 10s complement for the first time and for every successive column with a * mark we will take the 9s complement. This is the minor difference we were talking about.

We perform the following steps:
- 8 – 6 = 2
- 9s complement of 2 is 7 [9 – 2 = 7]

Write down 7 in the tens place of the result.

```
        *   *
    3   6   5
 -  1   8   6
 = _____
        7   9
```

Now we come to the 100s place. There is no * mark on top of the column. We love it when the number on the top is greater than the number at the bottom. Now, since we are coming out of complements, we will another adjustment to our subtraction process.

3 – 1 = 2 and 2 – 1 = 1; subtract 1 from the result to indicate that you have come out of complements or a set of columns with a * mark.

Write down 1 in the 100s place. Therefore the final answer is 179.

```
        *   *
    3   6   5
 -  1   8   6
 = _____
    1   7   9
```

For ease of explanation we will refer to a column with a * on top as a pattern, since it represents a pattern where the number on the top is lesser than the number at the bottom, or the first number is smaller than second.

So, the procedure can best be encapsulated by a simple rule of thumb.

1.  Identify patterns and mark them with *
2.  The first time you come across the pattern, subtract the smaller number from the larger (or subtract the first number from the second) and take the 10s complement. 10s complement is simply 10 minus the number.
3.  For every subsequent and consecutive occurrence of the pattern, we subtract the smaller number from larger (or subtract the first number from the second) and take 9s complement. 9s complement of a number is simply 9 minus the number.
4.  When the pattern is broken, subtract as always. And subtract one more from the result.

Let us solve one more problem to highlight this method. Once we get an understanding of the steps, we will see how easy it to actually subtract. The complexity is not in the steps. The complexity is in our attempt to explain the steps in English consistently.

Let us consider the following problem.

**Ex. 2**: 5322 − 3876

*Step 1*: Identify patterns and mark the columns with a *

```
          *   *   *
      5   3   2   2
  −   3   8   7   6
  ────────────────
  =
```

*Step 2*: Start with Units column on the right. We see the start of the pattern.

$6 - 2 = 4$; 10s complement of 4 is 6

```
          *   *   *
      5   3   2   2
  −   3   8   7   6
  ────────────────
  =               6
```

*Step 3*: We move to tens place. The '*' sign indicates that the pattern continues.

Therefore, 7 – 2 = 5; 9s complement of 5 is 4.

```
        *   *   *
    5   3   2   2
  - 3   8   7   6
  =         4   6
```

*Step 4*: We move to the 100s place. The '*' sign indicates that the pattern continues.

Therefore, 8 – 3 = 5; 9s complement of 5 is 4.

```
        *   *   *
    5   3   2   2
  - 3   8   7   6
  =     4   4   6
```

*Step 5*: We move to the leftmost column now. The * sign is absent. We are now coming out of the pattern and hence out of the complements.

Therefore, we perform 5 – 3 = 2; and 2 – 1 =1. We subtract 1 finally to indicate that we have come out of complements.

```
        *   *   *
    5   3   2   2
  - 3   8   7   6
  = 1   4   4   6
```

Therefore the answer is 1446.

At the end of the second example, some level of clarity has emerged. Let us solve another example, with a lot less amount of annotation and explanation. As you read through, you should try to solve the problem on a piece of paper to ensure that you have understood the steps.

**Ex. 3**: 4121 – 2687

| | | * | * | * | Step 1: Insert patterns with * |
|---|---|---|---|---|---|
| | 4 | 1 | 2 | 1 | |
| – | 2 | 6 | 8 | 7 | |
| | | | | 4 | Step 2: 7 – 1=6; 10s complement of 6 is 4 |
| | | | 3 | | Step 3: 8 – 2=6; 9s complement of 6 is 3 |
| | | 4 | | | Step 4: 6 – 1=5; 9s complement of 5 is 4 |
| | 1 | | | | Step 5: We are out of complements; 4 – 2=2; 2 – 1=1 |
| = | 1 | 4 | 3 | 4 | Final Result: 1434 |

The process of handling patterns is "all from 9 and last from 10". This means for the last column marked *, we take the 10s complement; for every successive column to its left marked with *, we take the 9s complement. We subtract 1 from the difference, when we come out of complements.

**Ex. 4**: 5432 –1867

| | | * | * | * | Step 1: Identify patterns with * |
|---|---|---|---|---|---|
| | 5 | 4 | 3 | 2 | |
| – | 1 | 8 | 6 | 7 | |
| | | | | 5 | Step 2: 7 – 2=5; 10s complement of 5 is 5 |
| | | | 6 | | Step 3: 6 – 3=3; 9s complement of 3 is 6 |
| | | 5 | | | Step 4: 8 – 4=4; 9s complement of 4 is 5 |
| | 3 | | | | Step 5: We are out of complements; 5 – 1=4; 4 – 1=3 |
| = | 3 | 5 | 6 | 5 | Final Result: 3565 |

**Ex. 5**: 6010 – 4872

Like before, we will capture the steps in a table and annotate the steps systematically to help you follow the sequence of operations.

| | | | | |
|---|---|---|---|---|
| | * | * | * | Step 1: Identify patterns with * |
| 6 | 0 | 1 | 0 | |
| – 4 | 8 | 7 | 2 | |
| | | | 8 | Step 2: 2 – 0=2; 10s complement of 2 is 8 |
| | | 3 | | Step 3: 7 – 1=6; 9s complement of 6 is 3 |
| | 1 | | | Step 4: 8 – 0=8; 9s complement of 8 is 1 |
| 1 | | | | Step 5: We are out of complements; 6 – 4=2; 2 – 1=1 |
| = 1 | 1 | 3 | 8 | Final Result: 1138 |

In the previous examples, we have added text to the table we used to solve the problem. With practice, you should be able to get these steps mentally. The detailed annotation and explanation makes it seems like a chore. However, when you get a handle of this solution process, you will simply proceed to the subtraction process without much ado.

Let us solve a few examples. We will proceed systematically; but you need to make an effort to recall the steps, because we will not be laying out the solution in a table as before. We will solve it just like we would during the course of a normal subtraction problem.

**Ex. 6**: 32487 – 14533
**Solution:**

```
      *  *
   3  2  4  8  7
 - 1  4  5  3  3
 = 1  7  9  5  4
```

**Ex. 7**: 76019 – 29122
**Solution:**

```
      *  *  *
   7  6  0  1  9
 - 2  9  1  2  2
 = 4  6  8  9  7
```

**Ex. 8**: 75642 – 19691
**Solution:**

```
        *   *   *
    7   5   6   4   2
-   1   9   6   9   1
=   5   5   9   5   1
```

**Ex. 9**: 13147 – 9456
**Solution:**

```
            *   *   *
    1   3   1   4   7
-       9   4   5   6
=   0   3   6   9   1
```

**Ex. 10**: 43720 – 9820
**Solution:**

```
        *   *
    4   3   7   2   0
-       9   8   2   0
=   3   3   9   0   0
```

**Ex. 11**: 73224 - 5862
**Solution:**

```
            *   *   *
    7   3   2   2   4
-       5   8   6   2
=   6   7   3   6   2
```

We have solved 6 examples so far to highlight the method. Let us spend a quick minute or two to recap the process and understand the salient features.

a. There is no borrowing step. Instead we handle the 10s and 9s complements.

b. There is no need for us to subtract a larger number from a smaller number.

We have thus come across a procedure for speed subtraction. And in this process, we have reduced the probability of error by eliminating the error prone steps and replacing them with operations that we are very good at.

## 4.1.1  Exercises for practice

**Prob. 1**: 5152 – 1763
**Solution:**

```
    5   1   5   2
-   1   7   6   3
=
```

**Prob. 2**: 3251 – 1762
**Solution:**

```
    3   2   5   1
-   1   7   6   2
=
```

**Prob. 3:** 4242 – 1453
**Solution:**

```
    4   2   4   2
–   1   4   5   3
=
```

**Prob. 4:** 5612 – 1877
**Solution:**

```
    5   6   1   2
–   1   8   7   7
=
```

**Prob. 5:** 34121 – 12678
**Solution:**

```
    3   2   1   2   1
–   1   2   6   7   8
=
```

**Prob. 6:** 35133 – 17249
**Solution:**

```
    3   5   1   3   3
–   1   7   2   4   9
=
```

**Prob. 7:** 8774 – 6666
**Solution:**

```
    8   7   7   4
–   6   6   6   6
=
```

**Prob. 8:** 56381 – 19670
**Solution:**

```
    5   6   3   8   1
–   1   9   6   7   0
=
```

**Prob. 9:** 4327 – 1514
**Solution:**

```
    4   3   2   7
–   1   5   1   4
=
```

**Prob. 10:** 3762 – 1980
**Solution:**

```
    3   7   6   2
–   1   9   8   0
=
```

**Prob. 11:** 646 – 171
**Solution:**

```
    6   4   6
–   1   7   1
=
```

**Prob. 12:** 32487 – 14533
**Solution:**

```
    3   2   4   8   7
–   1   4   5   3   3
=
```

**Prob. 13:** 43723 – 19782
**Solution:**

```
    4   3   7   2   3
–   1   9   7   8   2
=
```

**Prob. 14:** 32346 – 18212
**Solution:**

```
    3   2   3   4   6
–   1   8   2   1   2
=
```

**Prob. 15:** 64321 – 15431
**Solution:**

```
    6   4   3   2   1
–   1   5   4   3   1
=
```

**Prob. 16:** 66220 – 49120
**Solution:**

```
    6   6   2   2   0
–   4   9   1   2   0
=
```

**Prob. 17:** 7578 – 2967
**Solution:**

```
    7   5   7   8
–   2   9   6   7
=
```

**Prob. 18:** 5713 – 1246
**Solution:**

```
    5   7   1   3
–   1   2   4   6
=
```

**Prob. 19:** 6234 – 1078
**Solution:**

```
   6   2   3   4
-  1   0   7   8
=
```

**Prob. 20:** 7564 – 1298
**Solution:**

```
   7   5   6   4
-  1   2   9   8
=
```

**Prob. 21:** 2354 – 1068
**Solution:**

```
   2   3   5   4
-  1   0   6   8
=
```

**Prob. 22:** 4382 – 2794
**Solution:**

```
   4   3   8   2
-  2   7   9   4
=
```

Let us make a few more observations regarding complements. Let us consider 152 – 93 as the next example. In this example, let focus our attention on the tens place. We have 9–2=7; and we take the 9s complement. This means that we perform 9–7 and get 2. In other words, if the bottom number is 9 and we are under complements, we can safely write down the difference as the number on the top. 9–(9–2) is the same as 9–9+2; which leaves us with 2; which is the same as the number on the top.

If the second number is 8, the answer is first number+1; and so on. Identifying such patterns improves our speed of computation. Let us create a table to capture the number pattern that we just discussed.

| # | Second Number is | Final Answer |
|---|---|---|
| 1 | 9 | First Number |
| 2 | 8 | First Number+1 |
| 3 | 7 | First Number+2 |
| 4 | 6 | First Number + 3 |

There is a reason why we looked at this number pattern. Speed Math is simply a collection of useful number patterns that we deploy to solve problems without actual going through the grind of numerical operations.

Part of the expertise is related to our ability to detect such patterns and start playing with them.

What happens when patterns start and stop? Simple. We keep repeating the process that we learnt repeated.

**Ex. 12:** 50256 – 21078
**Solution:**

```
      *     * *
   5 0 2 5 6
 - 2 1 0 7 8
 = 2 9 1 7 8
```

In this example, we see the pattern train occurring two times. The basic problem solving process remains the same. The pattern starts in the units place; so we take the 10s complement of the difference of bottom row number and the top row number.

8 – 6 = 2; 10s complement of 2 is 8.

In the 10s place, we see the pattern continue. Here we take the 9s complement of 7 minus 5 which is 7. In the 100s place, the pattern ends. So we get, 2 minus 0 minus 1, which is 1.The pattern reappears in the next column, and we revert back to taking 10s complement.

**Ex. 13:** 21445 – 12399
**Solution:**

```
      *     * *
   2 1 4 4 5
 - 1 2 3 9 9
 = 0 9 0 4 6
```

**Ex. 14:** 54362 – 16178
**Solution:**

```
      *     * *
   5 4 3 6 2
 - 1 6 1 7 8
 = 3 8 1 8 4
```

Actually, you can start writing the difference from right to left, with a little practice!

**Ex. 15:** 74813 – 67404
**Solution:**

```
          *       *
    7   4   8   1   3
-   6   7   4   0   4
= 0   7   4   0   9
```

**Ex. 16:** 363239 – 177190
**Solution:**

```
          *   *       *
    3   6   3   2   3   9
-   1   7   7   1   9   0
= 1   8   6   0   4   9
```

We can now summarize the key steps.

1. Take the complement of the difference when the digit in the bottom row is greater than the digit in the top row.
2. The first is always 10s complement and the others are 9s complement
3. Come out of complements, when the digit in top row is greater than digit in bottom row.
4. When coming out of complements, we drop one more from the difference.

## 4.1.2  Exercises for practice

**Prob. 1:** 54326 - 12784
**Solution:**

```
    5   4   3   2   6
-   1   2   7   8   4
=
```

**Prob. 2:** 756432-345678
**Solution:**

```
    7   5   6   4   3   2
-   3   4   5   6   7   8
=
```

**Prob. 3:** 84612 - 72832
**Solution:**

```
    8  4  6  1  2
 -  7  2  8  3  2
 = _____
```

**Prob. 4:** 71209 - 34326
**Solution:**

```
    7  1  2  0  9
 -  3  4  3  2  6
 = _____
```

**Prob. 5:** 32654 - 12345
**Solution:**

```
    3  2  6  5  4
 -  1  2  3  4  5
 = _____
```

**Prob. 6:** 72306 - 51129
**Solution:**

```
    7  2  3  0  6
 -  5  1  1  2  9
 = _____
```

**Prob. 7:** 64156 - 42374
**Solution:**

```
    6  4  1  5  6
 -  4  2  3  7  4
 = _____
```

**Prob. 8:** 36271 - 12123
**Solution:**

```
    3  6  2  7  1
 -  1  2  1  2  3
 = _____
```

**Prob. 9:** 43215 - 18162
**Solution:**

```
    4  3  2  1  5
 -  1  8  1  6  2
 = _____
```

**Prob. 10:** 53408 - 20089
**Solution:**

```
    5  3  4  0  8
 -  2  0  0  8  9
 = _____
```

**Prob. 11:** 80000 - 76543
**Solution:**

```
    8  0  0  0  0
  − 7  6  5  4  3
  = _____
```

**Prob. 12:** 93264 - 17529
**Solution:**

```
    9  3  2  6  4
  − 1  7  5  2  9
  = _____
```

**Prob. 13:** 14562 - 8917
**Solution:**

```
    1  4  5  6  2
  −    8  9  1  7
  = _____
```

**Prob. 14:** 94563 - 45678
**Solution:**

```
    9  4  5  6  3
  − 4  5  6  7  8
  = _____
```

**Prob. 15:** 87654 - 69288
**Solution:**

```
    8  7  6  5  4
  − 6  9  2  8  8
  = _____
```

**Prob. 16:** 48374 - 29256
**Solution:**

```
    4  8  3  7  4
  − 2  9  2  5  6
  = _____
```

# 5 Multiplication without tables

The problem of multiplication can be represented by the following equation.

Multiplicand × Multiplier = Product

While dealing with Arithmetic, we know that the multiplier, multiplicand and the product are all numbers. The process of multiplication simply the process of multiplying the multiplier with the right-most digit of the multiplicand; writing doing the product and moving on to the digit to its immediate left. If this is confusing; let us consider an example.

12 × 3 = ?

We multiply 3 with 2 (the right most digit of the multiplicand 12). The product is 6. We write 6 as the right most digit of the final product. We then proceed to find the result of 3 × 1. We continue to do this until we complete accounting for the left-most digit of the multiplicand.

Thus we write 12 × 3 = 36

In the context of the speed math technique that we will be looking at in this section, we will refer to the digit of the multiplicand currently under consideration as the "*number*"; and the digit to its immediate right as its "*neighbor*". We have introduced a new labeling scheme for easing the way we look at the multiplication process.

Let us now consider the following example and read out the number and its neighbor.

Example:          123456739845798235
Number:          123456739845798235 (call out the current digit starting with the right-most digit)
Neighbor:        234567398457982350 (call out the *neighbor* of the digit to the right of the *number*; if there is no digit, treat it as zero)

The *neighbor* of 5 in the units place is treated as zero. Similarly, we can add a zero in front of the entire number. The *neighbor* of that zero is the left-most digit.

We will need to manipulate the *number* and its *neighbor* in order to complete the multiplication without tables. We will formulate new rules and techniques for the same.

## 5.1 The zero sandwich

We will always add a zero in front of the multiplicand. The value of the quantity does not change. This is because, 123 is equal to 0123. We will stop the multiplication process when we complete handling the sandwich number. This is a useful flag to ensure that we account for all digits in the multiplicand.

The concept is very simple. *Add a zero in front of the multiplicand before the process of multiplication begins.* As we start solving a few examples, this concept will become clear.

We will now give you an idea of how the rest of the chapter is organized.

## 5.2 Multiplication by 0

Any number multiplied by zero gives zero. It does not matter how large or small this number is. It does not matter how many digits are there in the multiplicand. If the multiplier is zero, then the product is zero.

Therefore,

| | | |
|---|---|---|
| $12345 \times 0$ | = | 0 |
| $128756 \times 0$ | = | 0 |
| $7378279 \times 0$ | = | 0 |

## 5.3 Multiplication by 1

Rule:
  1. *For every number, write the number.*

Since any number multiplied by 1 gives the number itself. In other words, if multiplier is 1, the product is the same as the multiplier.

Therefore,

| | | |
|---|---|---|
| 12345 × 1 | = | 12345 |
| 128756 × 1 | = | 128756 |
| 7378279 × 1 | = | 7378279 |

## 5.4  Multiplication by 2

Rule:
1. *Add a 0 in front of the multiplicand*
2. *For every number, double the number.*

**Ex. 1:** 5834 × 2
**Solution:**
Add a 0 in front of the multiplicand.

```
      0   5   8   3   4
×                     2
─────────────────────────
      0  ₁0  ₁6   6   8
=     1   1   6   6   8
```

**Ex. 2:** 89741 × 2
**Solution:**
Add a 0 in front of the multiplicand.

```
      0   8   9   7   4   1
×                         2
──────────────────────────────
      0  ₁6  ₁8  ₁4   8   2
=     1   7   9   4   8   2
```

**Ex. 3:** 54034 × 2
**Solution:**
Add a 0 in front of the multiplicand.

```
      0   5   4   0   3   4
×                         2
──────────────────────────────
         ₁0   8   0   6   8
=     1   0   8   0   6   8
```

**Ex. 4:** 3876 × 2
**Solution:**
Add a 0 in front of the multiplicand.

```
      0   3   8   7   6
×                     2
─────────────────────────
      0   6  ₁6  ₁4  ₁2
=         7   7   5   2
```

**Ex. 5:** 6880 × 2
**Solution:**
Add a 0 in front of the multiplicand.

**Ex. 6:** 10285 × 2
**Solution:**
Add a 0 in front of the multiplicand.

$$\begin{array}{rccccc} & 0 & 6 & 8 & 8 & 0 \\ \times & & & & & 2 \\ \hline & 0 & {}_1 2 & {}_1 6 & {}_1 6 & 0 \\ = & 1 & 3 & 7 & 6 & 0 \end{array}$$

$$\begin{array}{rcccccc} & 0 & 1 & 0 & 2 & 8 & 5 \\ \times & & & & & & 2 \\ \hline & & 2 & 0 & 4 & {}_1 6 & {}_1 0 \\ = & & 2 & 0 & 5 & 7 & 0 \end{array}$$

**Ex. 7:** $7184 \times 2$
**Solution:**
Add a 0 in front of the multiplicand.

$$\begin{array}{rccccc} & 0 & 7 & 1 & 8 & 4 \\ \times & & & & & 2 \\ \hline & 0 & {}_1 4 & 2 & {}_1 6 & 8 \\ = & 1 & 4 & 3 & 6 & 8 \end{array}$$

**Ex. 8:** $22584 \times 2$
**Solution:**
Add a 0 in front of the multiplicand.

$$\begin{array}{rcccccc} & 0 & 2 & 2 & 5 & 8 & 4 \\ \times & & & & & & 2 \\ \hline & 0 & 4 & 4 & {}_1 0 & {}_1 6 & 8 \\ = & 0 & 4 & 5 & 1 & 6 & 8 \end{array}$$

**Ex. 9:** $9654 \times 2$
**Solution:**
Add a 0 in front of the multiplicand.

$$\begin{array}{rccccc} & 0 & 9 & 6 & 5 & 4 \\ \times & & & & & 2 \\ \hline & & {}_1 8 & {}_1 2 & {}_1 0 & 8 \\ = & 1 & 9 & 3 & 0 & 8 \end{array}$$

**Ex. 10:** $678 \times 2$
**Solution:**
Add a 0 in front of the multiplicand.

$$\begin{array}{rcccc} & 0 & 6 & 7 & 8 \\ \times & & & & 2 \\ \hline & 0 & {}_1 2 & {}_1 4 & {}_1 6 \\ = & 1 & 3 & 5 & 6 \end{array}$$

## 5.4.1 Exercises for practice

**Prob. 1:** $8925 \times 2$
**Solution:**

$$\begin{array}{rccccc} & 0 & 8 & 9 & 2 & 5 \\ \times & & & & & 2 \\ \hline & & & & & \\ = & & & & & \end{array}$$

**Prob. 2:** $5428 \times 2$
**Solution:**

$$\begin{array}{rccccc} & 0 & 5 & 4 & 2 & 8 \\ \times & & & & & 2 \\ \hline & & & & & \\ = & & & & & \end{array}$$

**Prob. 3:** 5525 × 2
**Solution:**

|   | 0 | 5 | 5 | 2 | 5 |
|---|---|---|---|---|---|
| × |   |   |   |   | 2 |

=

**Prob. 4:** 858876 × 2
**Solution:**

|   | 0 | 8 | 5 | 8 | 8 | 7 | 6 |
|---|---|---|---|---|---|---|---|
| × |   |   |   |   |   |   | 2 |

=

**Prob. 5:** 998979 × 2
**Solution:**

|   | 0 | 9 | 9 | 8 | 9 | 7 | 9 |
|---|---|---|---|---|---|---|---|
| × |   |   |   |   |   |   | 2 |

=

**Prob. 6:** 787945 × 2
**Solution:**

|   | 0 | 7 | 8 | 7 | 9 | 4 | 5 |
|---|---|---|---|---|---|---|---|
| × |   |   |   |   |   |   | 2 |

=

## 5.5 Multiplication by 3

Rule:
1. *Add a 0 in front of the multiplicand*
2. *To number, add double the number*

**Ex. 1:** 38305 × 3
**Solution:**
Add a 0 in front of the multiplicand.

|   | 0 | 3 | 8 | 3 | 0 | 5 |
|---|---|---|---|---|---|---|
| × |   |   |   |   |   | 3 |
|   | 0 | 9 | $_2$4 | 9 | 0 | $_1$5 |
| = | 1 | 1 | 4 | 9 | 1 | 5 |

**Ex. 2:** 17743 × 3
**Solution:**
Add a 0 in front of the multiplicand.

|   | 0 | 1 | 7 | 7 | 4 | 3 |
|---|---|---|---|---|---|---|
| × |   |   |   |   |   | 3 |
|   | 0 | 3 | $_2$1 | $_2$1 | $_1$2 | 9 |
| = | 0 | 5 | 3 | 2 | 2 | 9 |

**Ex. 3:** 99129 × 3
**Solution:**
Add a 0 in front of the multiplicand.

**Ex. 4:** 62700 × 3
**Solution:**
Add a 0 in front of the multiplicand.

| 0 | 9 | 9 | 1 | 2 | 9 |
|---|---|---|---|---|---|
| × |   |   |   |   | 3 |
| 0 | $_2$7 | $_2$7 | 3 | 6 | $_2$7 |
| = 2 | 9 | 7 | 3 | 8 | 7 |

| 0 | 6 | 2 | 7 | 0 | 0 |
|---|---|---|---|---|---|
| × |   |   |   |   | 3 |
| 0 | $_1$8 | 6 | $_2$1 | 0 | 0 |
| = 1 | 8 | 8 | 1 | 0 | 0 |

**Ex. 5:** 54975 × 3
**Solution:**
Add a 0 in front of the multiplicand.

| 0 | 5 | 4 | 9 | 7 | 5 |
|---|---|---|---|---|---|
| × |   |   |   |   | 3 |
| 0 | $_1$5 | $_1$2 | $_2$7 | $_2$1 | $_1$5 |
| = 1 | 6 | 4 | 9 | 2 | 5 |

**Ex. 6:** 23174 × 3
**Solution:**
Add a 0 in front of the multiplicand.

| 0 | 2 | 3 | 1 | 7 | 4 |
|---|---|---|---|---|---|
| × |   |   |   |   | 3 |
| 0 | 6 | 9 | 3 | $_2$1 | $_1$2 |
| = 0 | 6 | 9 | 5 | 2 | 2 |

**Ex. 7:** 60635 × 3
**Solution:**
Add a 0 in front of the multiplicand.

| 0 | 6 | 0 | 6 | 3 | 5 |
|---|---|---|---|---|---|
| × |   |   |   |   | 3 |
| 0 | $_1$8 | 0 | $_1$8 | 9 | $_1$5 |
| = 1 | 8 | 1 | 9 | 0 | 5 |

**Ex. 8:** 59590 × 3
**Solution:**
Add a 0 in front of the multiplicand.

| 0 | 5 | 9 | 5 | 9 | 0 |
|---|---|---|---|---|---|
| × |   |   |   |   | 3 |
| 0 | $_1$5 | $_2$7 | $_1$5 | $_2$7 | 0 |
| = 1 | 7 | 8 | 7 | 7 | 0 |

**Ex. 9:** 89658 × 3
**Solution:**
Add a 0 in front of the multiplicand.

| 0 | 8 | 9 | 6 | 5 | 8 |
|---|---|---|---|---|---|
| × |   |   |   |   | 3 |
| 0 | $_2$4 | $_2$7 | $_1$8 | $_1$5 | $_2$4 |
| = 2 | 6 | 8 | 9 | 7 | 4 |

**Ex. 10:** 84278 × 3
**Solution:**
Add a 0 in front of the multiplicand.

| 0 | 8 | 4 | 2 | 7 | 8 |
|---|---|---|---|---|---|
| × |   |   |   |   | 3 |
| 0 | $_2$4 | $_1$2 | 6 | $_2$1 | $_2$4 |
| = 2 | 5 | 2 | 8 | 3 | 4 |

## 5.5.1 Exercises for practice

**Prob. 1:** 1794 × 3
**Solution:**

```
   0  1  7  9  4
×              3
─────────────────
=
```

**Prob. 2:** 7369 × 3
**Solution:**

```
   0  7  3  6  9
×              3
─────────────────
=
```

**Prob. 3:** 94767 × 3
**Solution:**

```
   0  9  4  7  6  7
×                 3
────────────────────
=
```

**Prob. 4:** 84960 × 3
**Solution:**

```
   0  8  4  9  6  0
×                 3
────────────────────
=
```

**Prob. 5:** 69718 × 3
**Solution:**

```
   0  6  9  7  1  8
×                 3
────────────────────
=
```

**Prob. 6:** 49901 × 3
**Solution:**

```
   0  4  9  9  0  1
×                 3
────────────────────
=
```

**Prob. 7:** 72039 × 3
**Solution:**

```
   0  7  2  0  3  9
×                 3
────────────────────
=
```

**Prob. 8:** 21667 × 3
**Solution:**

```
   0  2  1  6  6  7
×                 3
────────────────────
=
```

**Prob.** 9: 35630 × 3
**Solution:**

```
    0   3   5   6   3   0
×                       3
_____
=
```

**Prob.** 10: 10988 × 3
**Solution:**

```
    0   1   0   9   8   8
×                       3
_____
=
```

## 5.6 Multiplication by 4

Rule:

1. *Add a 0 in front of the multiplicand*
2. *Double the number; and double again*

**Ex.** 1: 74623 × 4
**Solution:**
Add a 0 in front of the multiplicand.

```
    0   7   4   6   2   3
×                       4
_____
    0  ₂8 ₁6 ₂4  8 ₁2
= 2   9   8   4   9   2
```

**Ex.** 2: 72616 × 4
**Solution:**
Add a 0 in front of the multiplicand.

```
    0   7   2   6   1   6
×                       4
_____
    0  ₂8  8 ₂4  4 ₂4
= 2   9   0   4   6   4
```

**Ex.** 3: 95194 × 4
**Solution:**
Add a 0 in front of the multiplicand.

```
    0   9   5   1   9   4
×                       4
_____
    0  ₃6 ₂0  4 ₃6 ₁6
= 3   8   0   7   7   6
```

**Ex.** 4: 43110 × 4
**Solution:**
Add a 0 in front of the multiplicand.

```
    0   4   3   1   1   0
×                       4
_____
    0  ₁6 ₁2  4   4   0
= 1   7   2   4   4   0
```

**Ex. 5:** $92343 \times 4$
**Solution:**
Add a 0 in front of the multi-plicand.

| | 0 | 9 | 2 | 3 | 4 | 3 |
|---|---|---|---|---|---|---|
| $\times$ | | | | | | 4 |
| | 0 | $_3$6 | 8 | $_1$2 | $_1$6 | $_1$2 |
| = | 3 | 6 | 9 | 3 | 7 | 2 |

**Ex. 6:** $62152 \times 4$
**Solution:**
Add a 0 in front of the mul-tiplicand.

| | 0 | 6 | 2 | 1 | 5 | 2 |
|---|---|---|---|---|---|---|
| $\times$ | | | | | | 4 |
| | 0 | $_2$4 | 8 | 4 | $_2$0 | 8 |
| = | 2 | 4 | 8 | 6 | 0 | 8 |

**Ex. 7:** $31647 \times 4$
**Solution:**
Add a 0 in front of the multi-plicand.

| | 0 | 3 | 1 | 6 | 4 | 7 |
|---|---|---|---|---|---|---|
| $\times$ | | | | | | 4 |
| | 0 | $_1$2 | 4 | $_2$4 | $_1$6 | $_2$8 |
| = | 1 | 2 | 6 | 5 | 8 | 8 |

**Ex. 8:** $16502 \times 4$
**Solution:**
Add a 0 in front of the mul-tiplicand.

| | 0 | 1 | 6 | 5 | 0 | 2 |
|---|---|---|---|---|---|---|
| $\times$ | | | | | | 4 |
| | 0 | 4 | $_2$4 | $_2$0 | 0 | 8 |
| = | 0 | 6 | 6 | 0 | 0 | 8 |

**Ex. 9:** $66748 \times 4$
**Solution:**
Add a 0 in front of the multi-plicand.

| | 0 | 6 | 6 | 7 | 4 | 8 |
|---|---|---|---|---|---|---|
| $\times$ | | | | | | 4 |
| | 0 | $_2$4 | $_2$4 | $_2$8 | $_1$6 | $_3$2 |
| = | 2 | 6 | 6 | 9 | 9 | 2 |

**Ex. 10:** $54616 \times 4$
**Solution:**
Add a 0 in front of the mul-tiplicand.

| | 0 | 5 | 4 | 6 | 1 | 6 |
|---|---|---|---|---|---|---|
| $\times$ | | | | | | 4 |
| | 0 | $_2$0 | $_1$6 | $_2$4 | 4 | $_2$4 |
| = | 2 | 1 | 8 | 4 | 6 | 4 |

## 5.6.1 Exercises for practice

**Prob.** 1: 64556 × 4
**Solution:**

```
    0  6  4  5  5  6
×                  4
─────────────────────

=
─────────────────────
```

**Prob.** 2: 50508 × 4
**Solution:**

```
    0  5  0  5  0  8
×                  4
─────────────────────

=
─────────────────────
```

**Prob.** 3: 81666 × 4
**Solution:**

```
    0  8  1  6  6  6
×                  4
─────────────────────

=
─────────────────────
```

**Prob.** 4: 26901 × 4
**Solution:**

```
    0  2  6  9  0  1
×                  4
─────────────────────

=
─────────────────────
```

**Prob.** 5: 67508 × 4
**Solution:**

```
    0  6  7  5  0  8
×                  4
─────────────────────

=
─────────────────────
```

**Prob.** 6: 37278 × 4
**Solution:**

```
    0  3  7  2  7  8
×                  4
─────────────────────

=
─────────────────────
```

**Prob.** 7: 52488 × 4
**Solution:**

```
    0  5  2  4  8  8
×                  4
─────────────────────

=
─────────────────────
```

**Prob.** 8: 30969 × 4
**Solution:**

```
    0  3  0  9  6  9
×                  4
─────────────────────

=
─────────────────────
```

**Prob. 9:** 85910 × 4
**Solution:**

```
    0  8  5  9  1  0
×                  4
_____

=
```

**Prob. 10:** 42672 × 4
**Solution:**

```
    0  4  2  6  7  4
×                  4
_____

=
```

## 5.7  Multiplication by 11

Rule:
1. *Add a 0 in front of the multiplicand*
2. *Bring down the last digit of the multiplicand as is*
3. *For every digit to its left, add the number to its neighbor*
4. *First number becomes the left hand number of the answer*

**Ex. 1:** 4236 × 11
**Solution:**
Add a 0 in front of the multiplicand.

| | | | | | |
|---|---|---|---|---|---|
| 0 | 4 | 2 | 3 | 6 | |
| × | | | | 1 | 1 |

| | | | | | | |
|---|---|---|---|---|---|---|
| | | | | | 6 | Step 1: Bring down the units digit 6 as it is |
| | | | | 9 | | Step 2: Add the number and neighbour: 3+6=9 |
| | | | 5 | | | Step 3: Add the number and neighbour: 2+3=5 |
| | | 6 | | | | Step 4: Add the number and neighbour: 4+2=6 |
| | 4 | | | | | Step 5: Add the number and neighbour: 0+4=4 |
| = | 4 | 6 | 5 | 9 | 6 | Final Answer |

This is why a 0 is added in front. It flags the end of the multiplication process.

**Ex. 2:** $633 \times 11$
**Solution:**
Add a 0 in front of the multiplicand.

| | 0 | 6 | 3 | 3 | |
|---|---|---|---|---|---|
| × | | | 1 | 1 | |
| | | | | 3 | Step 1: Bring down the units digit 3 as it is |
| | | | 6 | | Step 2: Add the number and neighbour: 3+3=6 |
| | | 9 | | | Step 3: Add the number and neighbour: 6+3=9 |
| | 6 | | | | Step 4: Add the number and neighbour: 0+6=6 |
| = | 6 | 9 | 6 | 3 | Final Answer |

Again, when you are clear, do these steps mentally and solve the subsequent problems. OK?

**Ex. 3:** $72324 \times 11$
**Solution:**
Add a 0 in front of the multiplicand.

```
    0  7  2  3  2  4
 ×              1  1
 _____
    7  9  5  5  6  4
 =  7  9  5  5  6  4
```

**Ex. 4:** $468 \times 11$
**Solution:**
Add a 0 in front of the multiplicand.

```
       0     4     6  8
 ×                 1  1
 _____
       4  ₁0  ₁4     8
 =     5     1     4  8
```

**Ex. 5:** $63452 \times 11$
**Solution:**
Add a 0 in front of the multiplicand.

```
    0  6  3  4  5  2
 ×              1  1
 _____
    6  9  7  9  7  2
 =  6  9  7  9  7  2
```

**Ex. 6:** $27153 \times 11$
**Solution:**
Add a 0 in front of the multiplicand.

```
    0  2  7  1  5  3
 ×              1  1
 _____
    2  9  8  6  8  3
 =  2  9  8  6  8  3
```

**Ex.** 7: 23518 × 11
**Solution:**
Add a 0 in front of the multiplicand.

```
    0  2  3  5  1  8
×               1  1
─────────────────────
    2  5  8  6  9  8
=   2  5  8  6  9  8
```

**Ex.** 8: 5645 × 11
**Solution:**
Add a 0 in front of the multiplicand.

```
    0  5  6  4  5
×            1  1
──────────────────
    5 ₁1 ₁0  9  5
=   6  2  0  9  5
```

**Ex.** 9: 65389 × 11
**Solution:**
Add a 0 in front of the multiplicand.

```
    0  6  5  3  8  9
×               1  1
─────────────────────
    6 ₁1  8 ₁1 ₁7  9
=   7  1  9  2  7  9
```

**Ex.** 10: 7856 × 11
**Solution:**
Add a 0 in front of the multiplicand.

```
    0  7  8  5  6
×            1  1
──────────────────
    7 ₁5 ₁3 ₁1  6
=   8  6  4  1  6
```

## 5.7.1 Exercises for practice

**Prob.** 1: 38291 × 11
**Solution:**

```
    0  3  8  2  9  1
×               1  1
─────────────────────

=
```

**Prob.** 2: 3882 × 11
**Solution:**

```
    0  3  8  8  2
×            1  1
──────────────────

=
```

**Prob.** 3: 127034 × 11
**Solution:**

```
    0  1  2  7  0  3  4
×                  1  1
────────────────────────

=
```

**Prob.** 4: 67854 × 11
**Solution:**

```
    0  6  7  8  5  4
×               1  1
─────────────────────

=
```

**Prob.** 5: 34685 × 11
**Solution:**

```
    0  3  4  6  8  5
×               1  1
─────────────────────

=  ─────────────────
```

**Prob.** 6: 89764 × 11
**Solution:**

```
    0  8  9  7  6  4
×               1  1
─────────────────────

=  ─────────────────
```

**Prob.** 7: 52623 × 11
**Solution:**

```
    0  5  2  6  2  3
×               1  1
─────────────────────

=  ─────────────────
```

**Prob.** 8: 41824 × 11
**Solution:**

```
    0  4  1  8  2  4
×               1  1
─────────────────────

=  ─────────────────
```

**Prob.** 9: 9087 × 11
**Solution:**

```
    0  9  0  8  7
×            1  1
───────────────────

=  ───────────────
```

**Prob.** 10: 1453 × 11
**Solution:**

```
    0  1  4  5  3
×            1  1
───────────────────

=  ───────────────
```

## 5.8  Multiplication by 12

Rule:
1. ***Add a 0 in front of the multiplicand***
2. ***Double the number and add the neighbor***

**Ex.** 1: 342 × 12
**Solution:**
Add a 0 in front of the multiplicand.

**Ex.** 2: 413 × 12
**Solution:**
Add a 0 in front of the multiplicand.

$$
\begin{array}{r}
0\ \ 3\ \ 4\ \ 2 \\
\times \qquad 1\ \ 2 \\
\hline
3\ \ {}_1 0\ \ {}_1 0\ \ 4 \\
= 4\ \ 1\ \ 0\ \ 4 \\
\hline
\end{array}
\qquad
\begin{array}{r}
0\ \ 4\ \ 1\ \ 3 \\
\times \qquad 1\ \ 2 \\
\hline
4\ \ 9\ \ 5\ \ 6 \\
= 4\ \ 9\ \ 5\ \ 6 \\
\hline
\end{array}
$$

**Ex. 3:** 5367 × 12
**Solution:**
Add a 0 in front of the multiplicand.

$$
\begin{array}{r}
0\ \ 5\ \ 3\ \ 6\ \ 7 \\
\times \qquad\quad 1\ \ 2 \\
\hline
5\ \ {}_1 3\ \ {}_1 2\ \ {}_1 9\ \ {}_1 4 \\
= 6\ \ 4\ \ 4\ \ 0\ \ 4 \\
\hline
\end{array}
$$

**Ex. 4:** 63457 × 12
**Solution:**
Add a 0 in front of the multiplicand.

$$
\begin{array}{r}
0\ \ 6\ \ 3\ \ 4\ \ 5\ \ 7 \\
\times \qquad\qquad 1\ \ 2 \\
\hline
6\ \ {}_1 5\ \ {}_1 0\ \ {}_1 3\ \ {}_1 7\ \ {}_1 4 \\
= 7\ \ 6\ \ 1\ \ 4\ \ 8\ \ 4 \\
\hline
\end{array}
$$

**Ex. 5:** 80541 × 12
**Solution:**
Add a 0 in front of the multiplicand.

$$
\begin{array}{r}
0\ \ 8\ \ 0\ \ 5\ \ 4\ \ 1 \\
\times \qquad\qquad 1\ \ 2 \\
\hline
8\ \ {}_1 6\ \ 5\ \ {}_1 4\ \ 9\ \ 2 \\
= 9\ \ 6\ \ 6\ \ 4\ \ 9\ \ 2 \\
\hline
\end{array}
$$

**Ex. 6:** 63247 × 12
**Solution:**
Add a 0 in front of the multiplicand.

$$
\begin{array}{r}
0\ \ 6\ \ 3\ \ 2\ \ 4\ \ 7 \\
\times \qquad\qquad 1\ \ 2 \\
\hline
6\ \ {}_1 5\ \ 8\ \ 8\ \ {}_1 5\ \ {}_1 4 \\
= 7\ \ 5\ \ 8\ \ 9\ \ 6\ \ 4 \\
\hline
\end{array}
$$

**Ex. 7:** 6589 × 12
**Solution:**
Add a 0 in front of the multiplicand.

$$
\begin{array}{r}
0\ \ 6\ \ 5\ \ 8\ \ 9 \\
\times \qquad\quad 1\ \ 2 \\
\hline
6\ \ {}_1 7\ \ {}_1 8\ \ {}_2 5\ \ {}_1 8 \\
= 7\ \ 9\ \ 0\ \ 6\ \ 8 \\
\hline
\end{array}
$$

**Ex. 8:** 4444 × 12
**Solution:**
Add a 0 in front of the multiplicand.

$$
\begin{array}{r}
0\ \ 4\ \ 4\ \ 4\ \ 4 \\
\times \qquad\quad 1\ \ 2 \\
\hline
4\ \ {}_1 2\ \ {}_1 2\ \ {}_1 2\ \ 8 \\
= 5\ \ 3\ \ 3\ \ 2\ \ 8 \\
\hline
\end{array}
$$

**Ex. 9:** 7349 × 12
**Solution:**
Add a 0 in front of the multiplicand.

```
    0   7   3   4   9
×               1   2
─────────────────────
    7  ₁7  ₁0  ₁7  ₁8
=   8   8   1   8   8
```

**Ex. 10:** 4008 × 12
**Solution:**
Add a 0 in front of the multiplicand.

```
    0   4   0   0   8
×               1   2
─────────────────────
    4   8   0   8  ₁6
=   4   8   0   9   6
```

## 5.8.1  Exercises for practice

**Prob. 1:** 32214 × 12
**Solution:**

```
    0   3   2   2   1   4
×                   1   2
─────────────────────────

=
```

**Prob. 2:** 5861 × 12
**Solution:**

```
    0   5   8   6   1
×               1   2
─────────────────────

=
```

**Prob. 3:** 7483 × 12
**Solution:**

```
    0   7   4   8   3
×               1   2
─────────────────────

=
```

**Prob. 4:** 30518 × 12
**Solution:**

```
    0   3   0   5   1   8
×                   1   2
─────────────────────────

=
```

**Prob. 5:** 41187 × 12
**Solution:**

```
    0   4   1   1   8   7
×                   1   2
─────────────────────────

=
```

**Prob. 6:** 64628 × 12
**Solution:**

```
    0   6   4   6   2   8
×                   1   2
─────────────────────────

=
```

**Prob.** 7: 9834 × 12
Solution:

```
    0  9  8  3  4
×              1  2
─────────────────────

=  ──────────────────
```

**Prob.** 8: 7483 × 12
Solution:

```
    0  7  4  8  3
×              1  2
─────────────────────

=  ──────────────────
```

**Prob.** 9: 305182 × 12
Solution:

```
    0  3  0  5  1  8  2
×                    1  2
────────────────────────

=  ─────────────────────
```

**Prob.** 10: 41187 × 12
Solution:

```
    0  4  1  1  8  7
×                 1  2
────────────────────────

=  ──────────────────
```

## 5.9 Multiplication by 5

Rule:

1. *Add a 0 in front of the multiplicand*
2. *Use half the neighbor; add 5 if the number is odd*

*Note:* We will never get carryovers, hence no need for the final simplification

**Ex.** 1: 5125 × 5
Solution:
Add a 0 in front of the multiplicand.

```
    0  5  1  2  5
×              5
─────────────────────
=   2  5  6  2  5
```

**Ex.** 2: 23181 × 5
Solution:
Add a 0 in front of the multiplicand.

```
    0  2  3  1  8  1
×                 5
─────────────────────
=   1  1  5  9  0  5
```

Basics of Speed Mathematics

**Ex. 3:** 6794 × 5
**Solution:**
Add a 0 in front of the multiplicand.

```
    0  6  7  9  4
×              5
= 3  3  9  7  0
```

**Ex. 4:** 85679 × 5
**Solution:**
Add a 0 in front of the multiplicand.

```
    0  8  5  6  7  9
×                 5
= 4  2  8  3  9  5
```

**Ex. 5:** 99776 × 5
**Solution:**
Add a 0 in front of the multiplicand.

```
    0  9  9  7  7  6
×                 5
= 4  9  8  8  8  0
```

**Ex. 6:** 83837 × 5
**Solution:**
Add a 0 in front of the multiplicand.

```
    0  8  3  8  3  7
×                 5
= 4  1  9  1  8  5
```

**Ex. 7:** 3831 × 5
**Solution:**
Add a 0 in front of the multiplicand.

```
    0  3  8  3  1
×              5
= 1  9  1  5  5
```

**Ex. 8:** 44641 × 5
**Solution:**
Add a 0 in front of the multiplicand.

```
    0  4  4  6  4  1
×                 5
= 2  2  3  2  0  5
```

**Ex. 9:** 78635 × 5
**Solution:**
Add a 0 in front of the multiplicand.

```
    0  7  8  6  3  5
×                 5
= 3  9  3  1  7  5
```

**Ex. 10:** 59801 × 5
**Solution:**
Add a 0 in front of the multiplicand.

```
    0  5  9  8  0  1
×                 5
= 2  9  9  0  0  5
```

## 5.9.1   Exercises for practice

**Prob.** 1: 92934 × 5
Solution:

```
    0  9  2  9  3  4
×                 5
=
```

**Prob.** 2: 8539 × 5
Solution:

```
    0  8  5  3  9
×              5
=
```

**Prob.** 3: 5678 × 5
Solution:

```
    0  5  6  7  8
×              5
=
```

**Prob.** 4: 45328 × 5
Solution:

```
    0  4  5  3  2  8
×                 5
=
```

**Prob.** 5: 63971 × 5
Solution:

```
    0  6  3  9  7  1
×                 5
=
```

**Prob.** 6: 7846 × 5
Solution:

```
    0  7  8  4  6
×              5
=
```

**Prob.** 7: 9273 × 5
Solution:

```
    0  9  2  7  3
×              5
=
```

**Prob.** 8: 87961 × 5
Solution:

```
    0  8  7  9  6  1
×                 5
=
```

**Prob.** 9: 98236 × 5
Solution:

```
    0  9  8  2  3  6
×                 5
=
```

**Prob.** 10: 36547 × 5
Solution:

```
    0  3  6  5  4  7
×                 5
=
```

# 5.10 Multiplication by 6

Rule:
1. *Add a 0 in front of the multiplicand*
2. *Add the number and half of its neighbor; add 5 more if the number is odd*

**Ex. 1:** 8988 × 6
**Solution:**
Add a 0 in front of the multiplicand.

```
    0   8   9   8   8
×                   6
─────────────────────
    4  ₁2  ₁8  ₁2   8
=   5   3   9   2   8
```

**Ex. 2:** 43654 × 6
**Solution:**
Add a 0 in front of the multiplicand.

```
    0   4   3   6   5   4
×                       6
─────────────────────────
    2   5  ₁1   8  ₁2   4
=   2   6   1   9   2   4
```

**Ex. 3:** 32867 × 6
**Solution:**
Add a 0 in front of the multiplicand.

```
    0   3   2   8   6   7
×                       6
─────────────────────────
    1   9   6  ₁1   9  ₁2
=   1   9   7   2   0   2
```

**Ex. 4:** 55667 × 6
**Solution:**
Add a 0 in front of the multiplicand.

```
    0   5   5   6   6   7
×                       6
─────────────────────────
    2  ₁2  ₁3   9   9  ₁2
=   3   3   4   0   0   2
```

**Ex. 5:** 80539 × 6
**Solution:**
Add a 0 in front of the multiplicand.

```
    0   8   0   5   3   9
×                       6
─────────────────────────
    4   8   2  ₁1  ₁2  ₁4
=   4   8   3   2   3   4
```

**Ex. 6:** 21550 × 6
**Solution:**
Add a 0 in front of the multiplicand.

```
    0   2   1   5   5   0
×                       6
─────────────────────────
    1   2   8  ₁2  ₁0   0
=   1   2   9   3   0   0
```

**Ex.** 7: 62284 × 6
**Solution:**
Add a 0 in front of the multiplicand.

```
    0   6   2   2   8   4
×                       6
───────────────────────────
    3   7   3   6  ₁0   4
=   3   7   3   7   0   4
```

**Ex.** 8: 5398 × 6
**Solution:**
Add a 0 in front of the multiplicand.

```
    0   5   3   9   8
×                   6
─────────────────────────
    2  ₁1  ₁2  ₁8   8
=   3   2   3   8   8
```

**Ex.** 9: 44203 × 6
**Solution:**
Add a 0 in front of the multiplicand.

```
    0   4   4   2   0   3
×                       6
───────────────────────────
    2   6   5   2   1   8
=   2   6   5   2   1   8
```

**Ex.** 10: 78943 × 6
**Solution:**
Add a 0 in front of the multiplicand.

```
    0   7   8   9   4   3
×                       6
───────────────────────────
    3  ₁6  ₁2  ₁6   5   8
=   4   7   3   6   5   8
```

## 5.10.1 Exercises for practice

**Prob.** 1: 91354 × 6
**Solution:**

```
    0   9   1   3   5   4
×                       6
───────────────────────────

=
```

**Prob.** 2: 8378 × 6
**Solution:**

```
    0   8   3   7   8
×                   6
─────────────────────────

=
```

**Prob.** 3: 9555 × 6
**Solution:**

```
    0   9   5   5   5
×                   6
─────────────────────────

=
```

**Prob.** 4: 3323 × 6
**Solution:**

```
    0   3   3   2   3
×                   6
─────────────────────────

=
```

**Prob.** 5: 44320 × 6
Solution:

```
  0  4  4  3  2  0
×              6
```

=

**Prob.** 6: 3333 × 6
Solution:

```
  0  3  3  3  3
×           6
```

=

**Prob.** 7: 67524 × 6
Solution:

```
  0  6  7  5  2  4
×              6
```

=

**Prob.** 8: 92341 × 6
Solution:

```
  0  9  2  3  4  1
×              6
```

=

**Prob.** 9: 56432 × 6
Solution:

```
  0  5  6  4  3  2
×              6
```

=

**Prob.** 10: 12345 × 6
Solution:

```
  0  1  2  3  4  5
×              6
```

=

# 5.11 Multiplication by 7

Rule:
1. *Add a 0 in front of the multiplicand*
2. *Double the number and add half of its neighbor; add 5 more if the number is odd*

**Ex.** 1: 63 × 7
Solution:
Add a 0 in front of the mul-
tiplicand.

```
    0  6  3
×        7
  ─────────
    3 ₁3 ₁1
=   4  4  1
```

**Ex.** 2: 458 × 7
Solution:
Add a 0 in front of the multi-
plicand.

```
    0  4  5  8
×           7
  ────────────
    2 ₁0 ₁9 ₁6
=   3  2  0  6
```

**Ex.** 3: 4038 × 7
**Solution:**
Add a 0 in front of the multiplicand.

$$\begin{array}{rccccc}
 & 0 & 4 & 0 & 3 & 8 \\
\times & & & & & 7 \\
\hline
 & 2 & 8 & 1 & {}_1 5 & {}_1 6 \\
= & 2 & 8 & 2 & 6 & 6
\end{array}$$

**Ex.** 4: 38579 × 7
**Solution:**
Add a 0 in front of the multiplicand.

$$\begin{array}{rcccccc}
 & 0 & 3 & 8 & 5 & 7 & 9 \\
\times & & & & & & 7 \\
\hline
 & 1 & {}_1 5 & {}_1 8 & {}_1 8 & {}_2 3 & {}_2 3 \\
= & 2 & 7 & 0 & 0 & 5 & 3
\end{array}$$

**Ex.** 5: 45328 × 7
**Solution:**
Add a 0 in front of the multiplicand.

$$\begin{array}{rcccccc}
 & 0 & 4 & 5 & 3 & 2 & 8 \\
\times & & & & & & 7 \\
\hline
 & 2 & {}_1 0 & {}_1 6 & {}_1 2 & 8 & {}_1 6 \\
= & 3 & 1 & 7 & 2 & 9 & 6
\end{array}$$

**Ex.** 6: 5072 × 7
**Solution:**
Add a 0 in front of the multiplicand.

$$\begin{array}{rcccccc}
 & 0 & 5 & 0 & 7 & 2 \\
\times & & & & & 7 \\
\hline
 & 2 & {}_1 5 & 3 & {}_2 0 & 4 \\
= & 3 & 5 & 5 & 0 & 4
\end{array}$$

**Ex.** 7: 695 × 7
**Solution:**
Add a 0 in front of the multiplicand.

$$\begin{array}{rcccc}
 & 0 & 6 & 9 & 5 \\
\times & & & & 7 \\
\hline
 & 3 & {}_1 6 & {}_2 5 & {}_1 5 \\
= & 4 & 8 & 6 & 5
\end{array}$$

**Ex.** 8: 2451 × 7
**Solution:**
Add a 0 in front of the multiplicand.

$$\begin{array}{rccccc}
 & 0 & 2 & 4 & 5 & 1 \\
\times & & & & & 7 \\
\hline
 & 1 & 6 & {}_1 0 & {}_1 5 & 7 \\
= & 1 & 7 & 1 & 5 & 7
\end{array}$$

**Ex.** 9: 8150 × 7
**Solution:**
Add a 0 in front of the multiplicand.

**Ex.** 10: 83564 × 7
**Solution:**
Add a 0 in front of the multiplicand.

```
    0  8  1  5  0                    0   8    3    5    6   4
×              7                  ×                        7
    4 ₁6  9 ₁5  0                   4  ₁7   ₁3   ₁8   ₁4   8
=   5  7  0  5  0                 =  5    8    4    9    4   8
```

## 5.11.1 Exercises for practice

**Prob.** 1: 7767 × 7
Solution:
```
    0  7  7  6  7
×              7
=
```

**Prob.** 2: 92081 × 7
Solution:
```
    0  9  2  0  8  1
×                 7
=
```

**Prob.** 3: 8074 × 7
Solution:
```
    0  8  0  7  4
×              7
=
```

**Prob.** 4: 52867 × 7
Solution:
```
    0  5  2  8  6  7
×                 7
=
```

**Prob.** 5: 9876 × 7
Solution:
```
    0  9  8  7  6
×              7
=
```

**Prob.** 6: 67592 × 7
Solution:
```
    0  6  7  5  9  2
×                 7
=
```

**Prob.** 7: 2738 × 7
Solution:
```
    0  2  7  3  8
×              7
=
```

**Prob.** 8: 13247 × 7
Solution:
```
    0  1  3  2  4  7
×                 7
=
```

**Prob.** 9: 72653 × 7
**Solution:**

```
  0  7  2  6  5  3
×              7
```

```
=
```

**Prob.** 10: 56432 × 7
**Solution:**

```
  0  5  6  4  3  2
×              7
```

```
=
```

## 5.12 Multiplication by 9

Rule:

1. *Add a 0 in front of the multiplicand*
2. *Subtract the right hand figure of the long number from ten. This gives the right-hand figure of the answer.*
3. *Taking each of the following figures in turn, up to the last one, subtract it from nine and add the neighbor.*
4. *At the last step when you are under the zero in front of the multiplicand, subtract one from the neighbor and use that as the left hand figure of the answer.*

**Ex.** 1: 8769 × 9
**Solution:**
Add a 0 in front of the multiplicand.

```
  0  8  7   6  9
×             9
  7  8  8  ₁2  1
= 7  8  9   2  1
```

**Ex.** 2: 8888 × 9
**Solution:**
Add a 0 in front of the multiplicand.

```
  0  8  8  8  8
×            9
  7  9  9  9  2
= 7  9  9  9  2
```

**Ex.** 3: 54331 × 9
**Solution:**
Add a 0 in front of the multiplicand.

**Ex.** 4: 6584 × 9
**Solution:**
Add a 0 in front of the multiplicand.

```
    0  5  4  3  3  1              0  6   5  8  4
×                 9          ×                9
    4  8  8  9  7  9              5  8  ₁2  5  6
=   4  8  8  9  7  9          =   5  9   2  5  6
```

**Ex. 5:** 7856 × 9
**Solution:**
Add a 0 in front of the multiplicand.

```
    0   7  8   5  6
×                9
    6  ₁0  6  ₁0  4
=   7   0  7   0  4
```

**Ex. 6:** 554 × 9
**Solution:**
Add a 0 in front of the multiplicand.

```
    0  5  5  4
×             9
    4  9  8  6
=   4  9  8  6
```

**Ex. 7:** 80532 × 9
**Solution:**
Add a 0 in front of the multiplicand.

```
    0  8   0  5  3  2
×                   9
    7  1  ₁4  7  8  8
=   7  2   4  7  8  8
```

**Ex. 8:** 53206 × 9
**Solution:**
Add a 0 in front of the multiplicand.

```
    0  5  3  2   0  6
×                   9
    4  7  8  7  ₁5  4
=   4  7  8  8   5  4
```

**Ex. 9:** 66381 × 9
**Solution:**
Add a 0 in front of the multiplicand.

```
    0  6  6   3  8  1
×                   9
    5  9  6  ₁4  2  9
=   5  9  7   4  2  9
```

**Ex. 10:** 53267 × 9
**Solution:**
Add a 0 in front of the multiplicand.

```
    0  5  3   2   6  7
×                    9
    4  7  8  ₁3  ₁0  3
=   4  7  9   4   0  3
```

## 5.12.1 Exercises for practice

**Prob.** 1: 6723 × 9
Solution:

```
  0  6  7  2  3
×             9
```

=

**Prob.** 2: 5897 × 9
Solution:

```
  0  5  8  9  7
×             9
```

=

**Prob.** 3: 7998 × 9
Solution:

```
  0  7  9  9  8
×             9
```

=

**Prob.** 4: 42365 × 9
Solution:

```
  0  4  2  3  6  5
×                9
```

=

**Prob.** 5: 5318 × 9
Solution:

```
  0  5  3  1  8
×             9
```

=

**Prob.** 6: 28222 × 9
Solution:

```
  0  2  8  2  2  2
×                9
```

=

**Prob.** 7: 23789 × 9
Solution:

```
  0  2  3  7  8  9
×                9
```

=

**Prob.** 8: 98456 × 9
Solution:

```
  0  9  8  4  5  6
×                9
```

=

**Prob.** 9: 672399 × 9
**Solution:**

$$0 \quad 6 \quad 7 \quad 2 \quad 3 \quad 9 \quad 9$$
$$\times \qquad\qquad\qquad\qquad 9$$
$$= \qquad\qquad\qquad\qquad\qquad$$

**Prob.** 10: 458971 × 9
**Solution:**

$$0 \quad 4 \quad 5 \quad 8 \quad 9 \quad 7 \quad 1$$
$$\times \qquad\qquad\qquad\qquad 9$$
$$= \qquad\qquad\qquad\qquad\qquad$$

# 5.13 Multiplication by 8

Rule:
1. *Add a 0 in front of the multiplicand*
2. *Subtract the right-hand digit from 10 and double*
3. *Subtract the middle figures from 9 and double then add with its neighbor*
4. *Subtract two from the neighbor of the last digit*

**Ex.** 1: 5905 × 8
**Solution:**
Add a 0 in front of the multiplicand.

$$0 \quad 5 \quad 9 \quad 0 \quad 5$$
$$\times \qquad\qquad\qquad 8$$
$$3 \quad {}_17 \quad 0 \quad {}_23 \quad {}_10$$
$$= 4 \quad 7 \quad 2 \quad 4 \quad 0$$

**Ex.** 2: 85236 × 8
**Solution:**
Add a 0 in front of the multiplicand.

$$0 \quad 8 \quad 5 \quad 2 \quad 3 \quad 6$$
$$\times \qquad\qquad\qquad\qquad 8$$
$$6 \quad 7 \quad {}_10 \quad {}_17 \quad {}_18 \quad 8$$
$$= 6 \quad 8 \quad 1 \quad 8 \quad 8 \quad 8$$

**Ex.** 3: 3214 × 8
**Solution:**
Add a 0 in front of the multiplicand.

$$0 \quad 3 \quad 2 \quad 1 \quad 4$$
$$\times \qquad\qquad\qquad 8$$
$$1 \quad {}_14 \quad {}_15 \quad {}_20 \quad {}_12$$
$$= 2 \quad 5 \quad 7 \quad 1 \quad 2$$

**Ex.** 4: 27903 × 8
**Solution:**
Add a 0 in front of the multiplicand.

$$0 \quad 2 \quad 7 \quad 9 \quad 0 \quad 3$$
$$\times \qquad\qquad\qquad\qquad 8$$
$$0 \quad {}_21 \quad {}_13 \quad 0 \quad {}_21 \quad {}_14$$
$$= 2 \quad 2 \quad 3 \quad 2 \quad 2 \quad 4$$

**Ex. 5:** 55339 × 8
**Solution:**
Add a 0 in front of the multiplicand.

```
    0   5   5   3   3   9
×                       8
    3  ₁3  ₁1  ₁5  ₂1   2
=   4   4   2   7   1   2
```

**Ex. 6:** 98731 × 8
**Solution:**
Add a 0 in front of the multiplicand.

```
    0   9   8   7   3   1
×                       8
    7   8   9   7  ₁3  ₁8
=   7   8   9   8   4   8
```

**Ex. 7:** 436 × 8
**Solution:**
Add a 0 in front of the multiplicand.

```
    0   4   3   6
×               8
    2  ₁3  ₁8   8
=   3   4   8   8
```

**Ex. 8:** 3323 × 8
**Solution:**
Add a 0 in front of the multiplicand.

```
    0   3   3   2   3
×                   8
    1  ₁5  ₁4  ₁7  ₁4
=   2   6   5   8   4
```

**Ex. 9:** 4567 × 8
**Solution:**
Add a 0 in front of the multiplicand.

```
    0   4   5   6   7
×                   8
    2  ₁5  ₁4  ₁3   6
=   3   6   5   3   6
```

**Ex. 10:** 98234 × 8
**Solution:**
Add a 0 in front of the multiplicand.

```
    0   9   8   2   3   4
×                       8
    7   8   4  ₁7  ₁6  ₁2
=   7   8   5   8   7   2
```

## 5.13.1 Exercises for practice

**Prob.** 1: 58631 × 8
Solution:

```
  0  5  8  6  3  1
×                 8
_____
=
```

**Prob.** 2: 4999 × 8
Solution:

```
  0  4  9  9  9
×              8
_____
=
```

**Prob.** 3: 1568 × 8
Solution:

```
  0  1  5  6  8
×              8
_____
=
```

**Prob.** 4: 66971 × 8
Solution:

```
  0  6  6  9  7  1
×                 8
_____
=
```

**Prob.** 5: 7755 × 8
Solution:

```
  0  7  7  5  5
×              8
_____
=
```

**Prob.** 6: 99883 × 8
Solution:

```
  0  9  9  8  8  3
x                 8
_____
=
```

**Prob.** 7: 58582 × 8
Solution:

```
  0  5  8  5  8  2
×                 8
_____
=
```

**Prob.** 8: 79902 × 8
Solution:

```
  0  7  9  9  0  2
×                 8
_____
=
```

**Prob.** 9: 38213 × 8
**Solution:**

    0  3  8  2  1  3

×                8

= _____

**Prob.** 10: 23953 × 8
**Solution:**

    0  2  3  9  5  3

×                8

= _____

## 5.14 Summary of the techniques for multiplying without tables

1. *Number: Any digit 0,1,2,3,4,5,6,7,8,9 or anything made up of these digits are called numbers*
2. *Neighbor: Any digit adjacent to a number, is called a neighbor*
3. *Half: Taking half of a digit, that is the value we get when dividing a digit by 2*
4. *Multiplication by 11:*
    a. *"Add each digit to its neighbor"*
5. *Multiplication by 12:*
    a. *"Double each digit and add its neighbor"*
6. *Multiplication by 5:*
    a. *"Use half the neighbor plus 5 if the number is odd"*
7. *Multiplication by 6:*
    a. *"To each number add half the neighbor plus 5 if the number is odd"*
8. *Multiplication by 7:*
    a. *"Double the number and add half the neighbor plus 5 if the number is odd"*
9. *Multiplication by 9:*
    a. *Subtract the right-hand digit from 10 and double*
    b. *Subtract the middle figures from 9 and double then add with its neighbor*
    c. *Subtract two from the neighbor of the last digit*
10. *Multiplication by 8:*
    a. *Subtract the right hand figure of the long number from ten. This gives the right-hand figure of the answer.*
    b. *Taking each of the following figures in turn, up to the last one, subtract it from nine and add the neighbor.*

      c.    *At the last step when you are under the zero in front of the multiplicand, subtract one from the neighbor and use that as the left hand figure of the answer*

11. *Multiplication by 4:*
      a.    *Double the number and double again*

12. *Multiplication by 3:*
      a.    *Double the number and add the number*

# 6 Base Multiplication

After a chapter on introductory topics, terms and terminology, we covered or rather uncovered the concepts in speed addition and subtraction. Now, we are ready to look at the crown jewel of speed math – the topic of speed multiplication. No matter what school of thought you consider, Abacus, Vedic Math, Trachtenberg or whatever the origin of techniques, multiplication is one topic that has attracted most attention. Therefore, there are several methods and techniques of attacking this problem space – faster and smarter ways for handling multiplication.

Most of us have trouble in handling multiplication. This is not because we are incompetent. It is because we have based this on rote learning of tables – without spending time on understanding the underlying beauty and patterns. We will proceed through this chapter very slowly. We will identify patterns and apply them to a few problems. These patterns will require a slight extension or modification to handle a class of numbers. Eventually we will attempt to cover all the possible scenarios.

Multiplication is simply shorthand for repeated addition.
$$3 + 3 + 3 + 3 = 3 \times 4$$

This means that three is added to itself four times. Therefore, we write the shorthand notation as $3 \times 4$ and read it as 3 times 4.

Let us now try and recall the concept of a base that we introduced in an earlier chapter. For getting a clear idea of the term 'base', we will solve a simple problem in multiplication.

**Ex. 1:** $9 \times 7$
**Solution:**

The idea is to use small numbers and easy examples so we focus our attention on the technique. What is the multiple of 10 closest to these numbers? 10.

We will now determine the deficiency of 9 and 7 from 10. This means we want to find 10 minus how much is 9? It is 1. Similarly, 10 minus how much is 7? 3.

We write down the problem as below:
Reference Number = 10

|  | Number | Sign | Deficiency |
|---|---|---|---|
|  | 9 | – | 1 |
| × | 7 | – | 3 |

The sign – indicates that 9 and 7 are short of the reference number 10.

Now we will split our product into parts. In other words,
$$9 \times 7 = \{part\text{-}1\} \ \{part\text{-}2\}$$

Let us now go on and solve the multiplication problem.

|  | Number | Sign | Deficiency |
|---|---|---|---|
|  | 9 | – | 1 |
| × | 7 | – | 3 |
| = | Part-1 |  | Part-2 |

We cross subtract the number-1 with the deficiency of number-2 or number-2 with deficiency of number-1. This gives us part-1 of the product.

Therefore, part-1 = 9 – 3 or 7 – 1 = 6

Part-2 of the product is simply the product of the deficiencies. In other words, part-2 = 1 × 3 = 3.

Therefore, 9 × 7 = {part-1} {part-2} = 6 3

Let us consider another example.
**Ex. 2:** 9 × 9
**Solution:**
We write down the reference number, which is 10.

|  | Number | Sign | Deficiency |
|---|---|---|---|
|  | 9 | – | 1 |
| × | 9 | – | 1 |
|  | (9 – 1) |  | (1 × 1) |
|  | 8 |  | 1 |
| = |  |  | 81 |

Let us try one more example.

**Ex.** 3: $9 \times 8$
**Solution:** Reference number = 10

| | Number | Sign | Deficiency |
|---|---|---|---|
| | 9 | − | 1 |
| × | 8 | − | 2 |
| | $(9-2)$ | | $(1 \times 2)$ |
| | 7 | | 2 |
| = | | | 72 |

The reference number 10 is also called the base. We will use the term "base" henceforth.

At this point, we are focusing on multiplication of single digit numbers that are below the base.

**Ex.** 4: $9 \times 6$
**Solution:** Base = 10

| | Number | Sign | Deficiency |
|---|---|---|---|
| | 9 | − | 1 |
| × | 6 | − | 4 |
| | $(9-4)$ | | $(1 \times 4)$ |
| | 5 | | 4 |
| = | | | 54 |

**Ex.** 5: $8 \times 7$
**Solution:** Base = 10

| | Number | Sign | Deficiency |
|---|---|---|---|
| | 8 | − | 2 |
| × | 7 | − | 3 |
| | $(8-3)$ | | $(2 \times 3)$ |
| | 5 | | 6 |
| = | | | 56 |

## 6.1.1 Exercises for practice

**Prob. 1:** 9 × 5
**Solution:**

| Number | Sign | Deficiency |
|---|---|---|
| 9 | – | |
| × 5 | – | |
| ( – ) | | ( × ) |

=

**Prob. 2:** 8 × 5
**Solution:**

| Number | Sign | Deficiency |
|---|---|---|
| 8 | – | |
| × 5 | – | |
| ( – ) | | ( × ) |

=

**Prob. 3:** 8 × 6
**Solution:**

| Number | Sign | Deficiency |
|---|---|---|
| 8 | – | |
| × 6 | – | |
| ( – ) | | ( × ) |

=

**Prob. 4:** 7 × 7
**Solution:**

| Number | Sign | Deficiency |
|---|---|---|
| 7 | – | |
| × 7 | – | |
| ( – ) | | ( × ) |

=

**Prob. 5:** 7 × 6
**Solution:**

| Number | Sign | Deficiency |
|---|---|---|
| 7 | – | |
| × 6 | – | |
| ( – ) | | ( × ) |

=

**Prob. 6:** 7 × 5
**Solution:**

| Number | Sign | Deficiency |
|---|---|---|
| 7 | – | |
| × 5 | – | |
| ( – ) | | ( × ) |

=

**Prob.** 7: 6 × 5
**Solution:**

| Number | Sign | Deficiency |
|---|---|---|
| 6 | − | |
| × 5 | − | |
| (−) | | (×) |
| = | | |

**Prob.** 8: 6 × 6
**Solution:**

| Number | Sign | Deficiency |
|---|---|---|
| 6 | − | |
| × 6 | − | |
| (−) | | (×) |
| = | | |

Let us see what happens when we try to use this method for determining 5 × 5.

**Ex. 6:** 5 × 5
**Solution:**

| Number | Sign | Deficiency |
|---|---|---|
| 5 | − | 5 |
| × 5 | − | 5 |
| (5 − 5) | | (5 × 5) |
| = 0 | | 5 × 5 |

In other words, this method expects you to know the product of 5 × 5. The product of the deficiency is the same the answer to the original problem. Clearly, we need extensions to this technique in order to handle this case. Or we need an alternative method for determining the value of 5 × 5. We will look at this case in a subsequent chapter.

Such limitations of patterns lead us to looking for ways to handle all sets of numbers. Like we said in the introduction to this chapter, we will introduce several techniques of multiplication. We need to make an informed choice on what works best and best for us. Let us summarize the key patterns.

When the number is below the base:
1. A multiplication operation can be replaced by a subtraction operation and a multiplication of small digits.
2. As the numbers move away from the base, the deficiencies increase and the product of deficiencies become a large multiplication problem.

3. It gives a general idea of a method to use to handle multiplication operation.
4. With this as the background, we now move on to handling four distinct classes of numbers, irrespective of the number of digits:
    a. Numbers close to and below the base
    b. Numbers close to and above the base
    c. Numbers close to and either side of the base
    d. Numbers away from the base

## 6.2 Numbers close to and below the base

Based on our definition of the term base, we can conclude that base for 2 digit numbers are 100, 3 digits numbers are 1000 and so on. Base of an n-digit number is simply 1 followed by n zeros. This is a critical concept for us to bear in mind as we start looking at product of large numbers which are close to the base number, but are both below the base number as well.

**Ex. 1:** $91 \times 91$
**Solution:** Base = 100
The process is identical to the one we used to handle the one digit case. We will now handle two digit numbers; therefore the base is one hundred. We compute deficiencies from 100 as well.

|  | Number | Sign | Deficiency |
|---|---|---|---|
|  | 91 | – | 09 |
| × | 91 | – | 09 |
|  | (91 – 09) |  | (09 × 09) |
|  | 82 |  | 81 |
| = |  |  | 8281 |

Part 1 of the product is: 91 - 09 = 82
Part 2 of the product is: 09 × 09 = 81
Part 1 tells us the number of Hundreds in the final product. 82 hundred is the same as 8200. Part 2 tells us the number of units in the final product. 81 units is the same as 8 Tens and 1 Unit.

Putting Part 1 next to Part 2, 81 next to 82 to make up 8281 is the same as adding 8200 + 81 = 8281. This is why looking at the final product in terms of parts makes a lot of sense. The process of addition of various parts is simply one of concatenation of the individual parts that make up the product.

Therefore, $91 \times 91 = 8281$

One thumb rule you want to remember is that deficiency must have as many digits as the number of 0s in the base. When base is equal to one hundred, we have two zeros. Therefore we write the deficiency as 09 and not 9.

**Ex. 2:** $89 \times 95$
**Solution:**

|   | Number | Sign | Deficiency |
|---|--------|------|------------|
|   | 89 | – | 11 |
| × | 95 | – | 05 |
|   | (89 – 05) |   | (11 × 05) |
|   | 84 |   | 55 |
| = |   |   | 8455 |

**Ex. 3:** $93 \times 93$
**Solution:**

|   | Number | Sign | Deficiency |
|---|--------|------|------------|
|   | 93 | – | 07 |
| × | 93 | – | 07 |
|   | (93 – 07) |   | (07 × 07) |
|   | 86 |   | 49 |
| = |   |   | 8649 |

**Ex. 4:** $93 \times 94$
**Solution:**

|   | Number | Sign | Deficiency |
|---|--------|------|------------|
|   | 93 | – | 07 |
| × | 94 | – | 06 |
|   | (93 – 06) |   | (07 × 06) |
|   | 87 |   | 42 |
| = |   |   | 8742 |

**Ex. 5:** $91 \times 96$
**Solution:**

|   | Number | Sign | Deficiency |
|---|--------|------|------------|
|   | 91 | – | 09 |
| × | 96 | – | 04 |
|   | (91 – 04) |   | (09 × 04) |
|   | 87 |   | 36 |
| = |   |   | 8736 |

**Ex. 6:** $93 \times 97$
**Solution:**

**Ex. 7:** $92 \times 98$
**Solution:**

| Number | Sign | Deficiency | | Number | Sign | Deficiency |
|---|---|---|---|---|---|---|
| 93 | – | 07 | | 92 | – | 08 |
| × 97 | – | 03 | × | 98 | – | 02 |
| (93 – 03) | | (07 × 03) | | (92 – 02) | | (08 × 02) |
| 90 | | 21 | | 90 | | 16 |
| = | | 9021 | = | | | 9016 |

**Ex. 8:** 98 × 88
**Solution:**

| Number | Sign | Deficiency |
|---|---|---|
| 88 | – | 12 |
| × 98 | – | 02 |
| (88 – 02) | | (12 × 02) |
| 86 | | 24 |
| = | | 8624 |

Although we were solving 98×88, we put 88 first and then 98. Considering the larger of the two numbers as the second number has its advantages. The larger number leaves a smaller deficiency. This makes computing the cross difference relatively easy and definitely error free. One can argue that 88 – 02 = 98 – 12 = 86. But 88 – 02 is easy, because the answer stares at our face. This will become clearer as we solve the next example.

**Ex. 9:** 78 × 97
**Solution:**

| Number | Sign | Deficiency |
|---|---|---|
| 78 | – | 22 |
| × 97 | – | 03 |
| (78 – 03) | | (22 × 03) |
| 75 | | 66 |
| = | | 7566 |

If we picked 97 – 22, we would have still come to 75. But 78 – 03 makes it error free. Subtracting a small number from another is easier.

**Ex.** 10: 88 × 96
**Solution:**

|   | Number | Sign | Deficiency |
|---|--------|------|------------|
|   | 88 | − | 12 |
| × | 96 | − | 04 |
|   | (88 − 04) |  | (12 × 04) |
|   | 84 |  | 48 |
| = |   |  | 8448 |

Therefore, the final product is 8448.

Let us now compute the deficiencies, cross subtract and consider the product of the deficiencies mentally. This section will give you a sense of true power of this speed multiplication technique.

This means that we will no longer use the grid. We will eliminate the prop. We will use the following thumb rule:
1. Part 1 is the "smaller number − deficiency of the larger number"; no matter what the original order of numbers is.
2. Part 2 is simply product of deficiencies.
3. We will separate the Part 1 and Part 2 with a space to highlight the two parts
4. We will finally write out our product.

For example,
**Ex. 11:** 88 × 96 = 88 − 04 / 12 × 04 = 84 48 = 8448
We look at the two numbers and mentally determine that the deficiencies are 12 and 04. Therefore the first part is 88 − 04; and second part is 12 × 04.

**Ex. 12:** 56 × 98 = 56 − 02 / 44 × 02 = 54 88 = 5488
In this problem, we can make an interesting observation. This method works very well − in terms of speed and accuracy - as long as one of the two numbers is close to the base. If we flipped the numbers the other way around, we would have spent a few more seconds on the cross difference.

**Ex. 13:** 98 × 56 = 98 − 44 / 02 × 44 = 54 88 = 5488
**Ex. 14:** 67 × 98 = 67 − 02 / 33 × 02 = 65 66 = 6566
**Ex. 15:** 25 × 99 = 25 − 01 / 75 × 01 = 24 75 = 2475

**Ex. 16:** $88 \times 88 = 88 - 12 / 12 \times 12 = 76\,_144 = 7744$
**Ex. 17:** $88 \times 91 = 88 - 09 / 12 \times 09 = 79\,_108 = 8008$
**Ex. 18:** $25 \times 98 = 25 - 02 / 75 \times 02 = 23\,_150 = 2450$
**Ex. 19:** $94 \times 98 = 94 - 02 / 06 \times 02 = 92\,12 = 9212$
**Ex. 20:** $98 \times 98 = 98 - 02 / 02 \times 02 = 96\,04 = 9604$
**Ex. 21:** $92 \times 99 = 92 - 01 / 08 \times 01 = 91\,08 = 9108$
**Ex. 22:** $94 \times 99 = 94 - 01 / 06 \times 01 = 93\,06 = 9306$
**Ex. 23:** $96 \times 99 = 96 - 01 / 04 \times 01 = 95\,04 = 9504$
**Ex. 24:** $98 \times 91 = 98 - 09 / 02 \times 09 = 89\,18 = 8918$
**Ex. 25:** $93 \times 91 = 93 - 09 / 07 \times 09 = 84\,63 = 8463$
**Ex. 26:** $93 \times 99 = 93 - 01 / 07 \times 01 = 92\,07 = 9207$

## 6.2.1 Exercises for practice

**Prob. 1:** $95 \times 97$
**Prob. 2:** $93 \times 98$
**Prob. 3:** $77 \times 98$
**Prob. 4:** $89 \times 97$
**Prob. 5:** $98 \times 81$
**Prob. 6:** $95 \times 98$
**Prob. 7:** $96 \times 98$
**Prob. 8:** $88 \times 96$
**Prob. 9:** $93 \times 96$
**Prob. 10:** $95 \times 99$
**Prob. 11:** $88 \times 92$
**Prob. 12:** $93 \times 92$
**Prob. 13:** $84 \times 95$
**Prob. 14:** $81 \times 97$
**Prob. 15:** $79 \times 96$

What happens when we have 3 digit numbers close to and below the base? Nothing changes. The strategy remains the same. The base simply increases to 1000 for 3 digit numbers, 10000 for 4 digit numbers and so on.

**Ex. 1:** $888 \times 998 = 888 - 002 / 112 \times 002 = 886\,224 = 886224$
**Ex. 2:** $988 \times 988 = 988 - 012 / 012 \times 012 = 976\,144 = 976144$
**Ex. 3:** $888 \times 991 = 888 - 009 / 112 \times 009 = 879\,_1008 = 880008$
**Ex. 4:** $998 \times 988 = 988 - 002 / 002 \times 012 = 986\,024 = 986024$
**Ex. 5:** $998 \times 997 = 998 - 003 / 002 \times 003 = 995\,006 = 995006$
**Ex. 6:** $994 \times 998 = 994 - 002 / 006 \times 002 = 992\,012 = 992012$

**Ex. 7:** $998 \times 998 = 998 - 002 / 002 \times 002 = 996\,004 = 996004$
**Ex. 8:** $992 \times 999 = 992 - 001 / 008 \times 001 = 991\,008 = 991008$
**Ex. 9:** $994 \times 999 = 994 - 001 / 006 \times 001 = 993\,006 = 993006$
**Ex. 10:** $996 \times 999 = 996 - 001 / 004 \times 001 = 995\,004 = 995004$
**Ex. 11:** $998 \times 991 = 991 - 002 / 009 \times 002 = 989\,018 = 989018$
**Ex. 12:** $993 \times 999 = 993 - 001 / 007 \times 001 = 992\,007 = 992007$

## 6.2.2 Exercises for practice

**Prob. 1:** $995 \times 997$
**Prob. 2:** $993 \times 998$
**Prob. 3:** $974 \times 998$
**Prob. 4:** $996 \times 997$
**Prob. 5:** $600 \times 998$
**Prob. 6:** $996 \times 900$
**Prob. 7:** $873 \times 997$
**Prob. 8:** $697 \times 897$
**Prob. 9:** $598 \times 998$
**Prob. 10:** $112 \times 998$
**Prob. 11:** $9997 \times 9997$
**Prob. 12:** $99979 \times 99999$
**Prob. 13:** $99988 \times 99997$

## 6.3 Numbers close to and above the base

We will now look at numbers close to and above the base. The process remains the same. We no longer have deficiencies. We have excess instead.

**Ex. 1:** $13 \times 12$
**Solution:** Base $= 10$

|   | Number | Sign | Excess |
|---|--------|------|--------|
|   | 13 | + | 3 |
| × | 12 | + | 2 |
|   | $(13+2)$ |  | $(3 \times 2)$ |
|   | 15 |  | 6 |
| = |  |  | 156 |

The + sign indicates the excess. We "cross add" instead of the cross subtract.

Part 1 of the product is 13 + 2 = 15; Part 2 is 3 × 2 = 6
Therefore the product is 156.

**Ex. 2:** 15 × 12
**Solution:** Base = 10

| | Number | Sign | Excess |
|---|---|---|---|
| | 15 | + | 5 |
| × | 12 | + | 2 |
| | (15 + 2) | | (5 × 2) |
| | 17 | | $_1$0 |
| = | | | 180 |

**Ex. 3:** 14 × 13
**Solution:** Base = 10

| | Number | Sign | Excess |
|---|---|---|---|
| | 14 | + | 4 |
| × | 13 | + | 3 |
| | (14 + 3) | | (4 × 3) |
| | 17 | | $_1$2 |
| = | | | 182 |

**Ex. 4:** 15 × 16
**Solution:** Base = 10

| | Number | Sign | Excess |
|---|---|---|---|
| | 15 | + | 5 |
| × | 16 | + | 6 |
| | (15 + 6) | | (5 × 6) |
| | 21 | | $_3$0 |
| = | | | 240 |

**Ex. 5:** 11 × 14
**Solution:** Base = 10

| | Number | Sign | Excess |
|---|---|---|---|
| | 11 | + | 1 |
| × | 14 | + | 4 |
| | (11 + 4) | | (1 × 4) |
| | 15 | | 4 |
| = | | | 154 |

**Ex. 6:** 12 × 18
**Solution:** Base = 10

| | Number | Sign | Excess |
|---|---|---|---|
| | 18 | + | 8 |
| × | 12 | + | 2 |
| | (18 + 2) | | (8 × 2) |
| | 20 | | $_1$6 |
| = | | | 216 |

Therefore the product is 216. Notice that we have used the larger of the two numbers first. This means that we need to add smaller excess from the smaller number. The choice is made based on a simple question: Which number gives me the lesser of the two deficiencies or excess?

**Ex.** 7: 101 × 103
**Solution:** Base = 100

| | Number | Sign | Excess |
|---|---|---|---|
| | 103 | + | 03 |
| × | 101 | + | 01 |
| | (103 + 01) | | (03 × 01) |
| | 104 | | 03 |
| = | | | 10403 |

Therefore the product is 10403. *The base is one hundred; therefore the excess must have two digits.*

**Ex.** 8: 101 × 101
**Solution:** Base = 100

| | Number | Sign | Excess |
|---|---|---|---|
| | 101 | + | 01 |
| × | 101 | + | 01 |
| | (101 + 01) | | (01 × 01) |
| | 102 | | 01 |
| = | | | 10201 |

**Ex.** 9: 102 × 104
**Solution:** Base = 100

| | Number | Sign | Excess |
|---|---|---|---|
| | 104 | + | 04 |
| × | 102 | + | 02 |
| | (104 + 02) | | (04 × 02) |
| | 106 | | 08 |
| = | | | 10608 |

**Ex.** 10: 110 × 102
**Solution:** Base = 100

| | Number | Sign | Excess |
|---|---|---|---|
| | 110 | + | 10 |
| × | 102 | + | 02 |
| | (110 + 02) | | (10 × 02) |
| | 112 | | 20 |
| = | | | 11220 |

**Ex.** 11: 105 × 109
**Solution:** Base = 100

| | Number | Sign | Excess |
|---|---|---|---|
| | 109 | + | 09 |
| × | 105 | + | 05 |
| | (109 + 05) | | (09 × 05) |
| | 114 | | 45 |
| = | | | 11445 |

In the next section, we will eliminate the grid and steps. We will compute the product mentally and lay them out systematically, like we did in the previous section.

Examples:
**Ex. 1:** $104 \times 109 = 109 + 04 / 09 \times 04 = 113\ 36 = 11336$
**Ex. 2:** $102 \times 115 = 115 + 02 / 15 \times 02 = 117\ 30 = 11730$
**Ex. 3:** $102 \times 165 = 165 + 02 / 65 \times 02 = 167\ _130 = 16830$
**Ex. 4:** $103 \times 120 = 120 + 03 / 20 \times 03 = 123\ 60 = 12360$
**Ex. 5:** $110 \times 121 = 121 + 10 / 21 \times 10 = 131\ _210 = 13310$
**Ex. 6:** $107 \times 110 = 110 + 07 / 10 \times 07 = 117\ 70 = 11770$
**Ex. 7:** $103 \times 112 = 112 + 03 / 12 \times 03 = 115\ 36 = 11536$
**Ex. 8:** $104 \times 118 = 118 + 04 / 18 \times 04 = 122\ 72 = 12272$
**Ex. 9:** $104 \times 112 = 112 + 04 / 12 \times 04 = 116\ 48 = 11648$
**Ex. 10:** $107 \times 109 = 109 + 07 / 09 \times 07 = 116\ 63 = 11663$

## 6.3.1 Exercises for practice

**Prob. 1:** $108 \times 106$
**Prob. 2:** $112 \times 113$
**Prob. 3:** $111 \times 116$
**Prob. 4:** $115 \times 115$
**Prob. 5:** $120 \times 104$
**Prob. 6:** $117 \times 103$
**Prob. 7:** $108 \times 112$
**Prob. 8:** $117 \times 112$
**Prob. 9:** $118 \times 111$
**Prob. 10:** $119 \times 115$

## 6.4 Numbers close to and on either side of the base

The basic process is similar to the techniques we have seen so far.

When the numbers are on the same side and below the base, the right side of the product was simply the product of the deficiency. The left side was the cross difference between the first number and the deficiency of the second.

For example, 98 × 98 was simply:

| | Number | Sign | Deficiency |
|---|---|---|---|
| | 98 | – | 02 |
| × | 98 | – | 02 |

Now, if we considered the negative sign along with the deficiency, we would get –02 × –02 = 04. This is the same as 02 × 02 = 04. This is the second part of the product. The first part is simply 98 – 02. In other words, we add the first number with the deficiency with the sign. This is simply, 98 + (–02) = 96.

When the numbers are on the same side and above the base, the right side of the product was the product of excess or deficiency. The left side was the cross addition of the first number and the excess of the second.

For example, 102 × 102 was simply:

| | Number | Sign | Excess |
|---|---|---|---|
| | 102 | + | 02 |
| × | 102 | + | 02 |

Like in the previous example, let us consider the sign of the excess. The second part of the product is simply +02 × +02 = 04. This is the same as 02 × 02 = 04; that we have been using in the previous section. The first part is the sum of the first number with the excess of the second with the sign included. This is the same as 102 + (+02) = 104.

This helps us to generalize these two methods into one.

In base multiplication:
1. The first part is simply the sum of the first number and the deficiency or excess with the sign.
2. The second part is simply the product of the excess or deficiency with the sign.

This generalization can help us simplify the problem solving process while handling numbers close to and either side of the base.

**Ex. 1:** 102 × 98
**Solution:** Base = 100

| | Number | Sign | Excess |
|---|---|---|---|
| | 98 | – | 02 |
| × | 102 | + | 02 |
| | (98 + 02) | | (–02 × +04) |
| | 100 | | –04 |
| = | 100$\overline{04}$ | = | 9996 |

Therefore the product is 100$\overline{04}$ = 9996

We could have completed the same this with (102 + –02) and (–02 × +02) for the first and second parts. The process will work fine. No issues. The key is to ensure that you use the sign in front of the excess or deficiency.

**Ex. 2:** 98 × 108
**Solution:** Base = 100

| | Number | Sign | Excess |
|---|---|---|---|
| | 98 | – | 02 |
| × | 108 | + | 08 |
| | (98 + 08) | | (–02 × 08) |
| | 106 | | –16 |
| = | 106$\overline{16}$ | = | 10584 |

**Ex. 3:** 81 × 104
**Solution:** Base = 100

| | Number | Sign | Excess |
|---|---|---|---|
| | 81 | – | 19 |
| × | 104 | + | 04 |
| | (81 + 04) | | (–04 × 19) |
| | 85 | | –76 |
| = | 85$\overline{76}$ | = | 8424 |

**Ex. 4:** 91 × 116
**Solution:** Base = 100

| | Number | Sign | Excess |
|---|---|---|---|
| | 91 | – | 09 |
| × | 116 | + | 16 |
| | (91 + 16) | | (–09 × 16) |
| | 107 | | –$_1$44 |
| = | 107$_1$44 | = | 10556 |

**Ex. 5:** 93 × 110
**Solution:** Base = 100

| | Number | Sign | Excess |
|---|---|---|---|
| | 93 | – | 07 |
| × | 110 | + | 10 |
| | (93 + 10) | | (–07 × 10) |
| | 103 | | –70 |
| = | 103$\overline{70}$ | = | 10230 |

**Ex. 6:** $88 \times 103$
**Solution:** Base = 100

| | Number | Sign | Excess |
|---|---|---|---|
| | 88 | − | 12 |
| × | 103 | + | 03 |
| | $(88 + 03)$ | | $(-12 \times 03)$ |
| | $\overline{91}$ | | −36 |
| = | 9136 | = | 9064 |

**Ex. 7:** $84 \times 120$
**Solution:** Base = 100

| | Number | Sign | Excess |
|---|---|---|---|
| | 84 | − | 16 |
| × | 120 | + | 20 |
| | $(84 + 20)$ | | $(-16 \times 20)$ |
| | 104 | | $-_3 20$ |
| = | $104_{\,3} 20$ | = | 10080 |

**Ex. 8:** $995 \times 1005$
**Solution:** Base = 1000

| | Number | Sign | Excess |
|---|---|---|---|
| | 995 | − | 5 |
| × | 1005 | + | 5 |
| | $(995 + 5)$ | | $(-5 \times 5)$ |
| | 1000 | | −025 |
| = | $100\overline{0025}$ | = | 999975 |

**Ex. 9:** $991 \times 1008$
**Solution:** Base = 1000

| | Number | Sign | Excess |
|---|---|---|---|
| | 991 | − | 9 |
| × | 1008 | + | 8 |
| | $(991 + 8)$ | | $(-9 \times 8)$ |
| | 999 | | −072 |
| = | $999\overline{072}$ | = | 998928 |

**Ex. 10:** $993 \times 1015$
**Solution:** Base = 1000

| | Number | Sign | Excess |
|---|---|---|---|
| | 993 | − | 7 |
| × | 1015 | + | 15 |
| | $(993 + 15)$ | | $(-7 \times 15)$ |
| | 1008 | | −105 |
| = | $1008\overline{105}$ | = | 1007895 |

**Ex. 11:** $983 \times 1016$
**Solution:** Base = 1000

| | Number | Sign | Excess |
|---|---|---|---|
| | 983 | − | 17 |
| × | 1016 | + | 16 |
| | $(983 + 16)$ | | $(-17 \times 16)$ |
| | 999 | | −272 |
| = | $999\overline{272}$ | = | 998728 |

**Ex. 12:** $84 \times 103 = 84 + 03 / -16 \times 03 = 87 / -48 = 8652$
**Ex. 13:** $97 \times 103 = 97 + 03 / -03 \times 03 = 100 / -09 = 9991$
**Ex. 14:** $91 \times 102 = 91 + 02 / -09 \times 02 = 93 / -18 = 9282$
**Ex. 15:** $88 \times 107 = 88 + 07 / -12 \times 07 = 95 / -84 = 9416$
**Ex. 16:** $92 \times 102 = 92 + 02 / -08 \times 02 = 94 / -16 = 9384$

**Ex. 17:** $81 \times 102 = 81 + 02 / -19 \times 02 = 83 / -38 = 8262$
**Ex. 18:** $98 \times 110 = 98 + 10 / -02 \times 10 = 108 / -20 = 10780$
**Ex. 19:** $83 \times 117 = 83 + 17 / -17 \times 17 = 100 / -289 = 9711$
**Ex. 20:** $93 \times 103 = 93 + 03 / -07 \times 03 = 96 / -21 = 9579$
**Ex. 21:** $86 \times 101 = 86 + 01 / -14 \times 01 = 87 / -14 = 8686$
**Ex. 22:** $889 \times 1011 = 889 + 011 / -111 \times 011 = 900 / -1221 = 898779$
**Ex. 23:** $985 \times 1005 = 985 + 005 / -015 \times 005 = 990 / -075 = 989925$
**Ex. 24:** $992 \times 1006 = 992 + 006 / -008 \times 006 = 998 / -048 = 997952$
**Ex. 25:** $996 \times 1030 = 996 + 030 / -004 \times 030 = 1026 / -120 = 1025880$
**Ex. 26:** $965 \times 1010 = 965 + 010 / -035 \times 010 = 975 / -350 = 974650$

### 6.4.1 Exercises for practice

**Prob. 1:** $94 \times 107$
**Prob. 2:** $94 \times 110$
**Prob. 3:** $96 \times 113$
**Prob. 4:** $83 \times 116$
**Prob. 5:** $89 \times 111$
**Prob. 6:** $991 \times 1050$
**Prob. 7:** $995 \times 1115$
**Prob. 8:** $849 \times 1006$
**Prob. 9:** $835 \times 1004$
**Prob. 10:** $997 \times 1025$
**Prob. 11:** $882 \times 1050$
**Prob. 12:** $897 \times 1060$
**Prob. 13:** $909 \times 1002$
**Prob. 14:** $867 \times 1040$

## 6.5 Numbers away from the base

When the numbers are far away from the base, we find that the cross-add and cross-subtract becomes cumbersome and the product of the deficiency or the excess is no longer a trivial problem.

For example, if we were trying to compute $63 \times 57$,
$$63 \times 57 = 63 - 43 / 37 \times 43 = 20 / 1591 = 3591$$

This example highlights the need for an alternative approach to handle this. One look at the two numbers tells us that the problem is easier to handle if the base was a number other than 100; let us say 60 in this case. The good news is that we can use a number other than 100 for a base. In fact, we can use 60 as the working base. This is how the solution space appears.

**Ex. 1:** 63 × 57
**Solution:**
Base = 10, Working base = 60 = 6 × 10, Factor = 6

| | Number | Sign | Excess |
|---|---|---|---|
| | 63 | + | 3 |
| × | 57 | − | 3 |
| | (63 − 3) | | (3 × -3) |
| | 60 | | −9 |

Deficiency and excess must be a single digit number, since base is 10. Now multiply the Part 1 of the product by the same factor by which you converted the base to a working base. In this example, Working Base (60) = 6 × 10 = Factor × Base (10).

$$60 \times 6 \mid -9$$
$$= \quad 360 \mid -9$$
$$= \quad 359 \mid 1$$

Therefore the product is 3591.

Let us use 100 as the base and see what happens. Now deficiency or excess must be a two digit number.
Base = 100, Working Base = 60 = 3/5 × 100, Factor = 3/5

| | Number | Sign | Excess |
|---|---|---|---|
| | 63 | + | 03 |
| × | 57 | − | 03 |
| | (63 − 03) | | (03 × -03) |
| | 60 | | −09 |
| | 60 × 3 / 5 | | −09 |
| | 36 | | −09 |
| = | 3609 | = | 3591 |

We arrived at the correct answer, by making two different choices for a base. The choice of the base impacts the number of digits in the deficiency/excess. If we make the right choice of the base, then we can apply this technique to numbers that are far away from the base. You can readily see that the base multiplication method can be extended to a variety of situations and it is very flexible.

**Ex. 2:** $61 \times 62$
Solution:
Base $= 60 = 6 \times 10$

| | Number | Sign | Excess |
|---|---|---|---|
| | 61 | + | 1 |
| × | 62 | + | 2 |
| | $(61+2)$ | | $(1\times2)$ |
| | 63 | | 2 |
| | $63\times6$ | | 2 |
| | 378 | | 2 |
| = | | | 3782 |

**Ex. 3:** $65 \times 68$
Solution:
Base $= 50 = 1/2 \times 100$

| | Number | Sign | Excess |
|---|---|---|---|
| | 65 | + | 15 |
| × | 68 | + | 18 |
| | $(65+18)$ | | $(15\times18)$ |
| | 83 | | 270 |
| | $83\times1/2$ | | 270 |
| | 41.5 | | 270 |
| = | $4150+270$ | = | 4420 |

**Ex. 4:** $55 \times 68$
Solution:
Base $= 50 = 1/2 \times 100$

| | Number | Sign | Excess |
|---|---|---|---|
| | 55 | + | 5 |
| × | 68 | + | 18 |
| | $(55+18)$ | | $(5\times18)$ |
| | 73 | | 90 |
| | $73\times1/2$ | | 90 |
| | 36.5 | | 90 |
| = | $3650+90$ | = | 3740 |

**Ex. 5:** $64 \times 54$
Solution:
Base $= 50 = 1/2 \times 100$

| | Number | Sign | Excess |
|---|---|---|---|
| | 64 | + | 14 |
| × | 54 | + | 4 |
| | $(64+4)$ | | $(14\times4)$ |
| | 68 | | 56 |
| | $68\times1/2$ | | 56 |
| | 34 | | 56 |
| = | $3400+56$ | = | 3456 |

**Ex. 6:** $67 \times 65$
Solution:
Base = $60 = 3/5 \times 100$

| | Number | Sign | Excess |
|---|---|---|---|
| | 67 | + | 7 |
| × | 65 | + | 5 |
| | $(67+5)$ | | $(7 \times 5)$ |
| | 72 | | 35 |
| | $72 \times 3/5$ | | 35 |
| | 43.2 | | 35 |
| = | $4320 + 35$ | = | 4355 |

**Ex. 7:** $60 \times 53$
Solution:
Base = $50 = 1/2 \times 100$

| | Number | Sign | Excess |
|---|---|---|---|
| | 60 | + | 10 |
| × | 53 | + | 3 |
| | $(60+3)$ | | $(10 \times 3)$ |
| | 63 | | 30 |
| | $63 \times 1/2$ | | 30 |
| | 31.5 | | 30 |
| = | $3150 + 30$ | = | 3180 |

**Ex. 8:** $62 \times 61$
**Solution:**
Base = $50 = 1/2 \times 100$

| | Number | Sign | Excess |
|---|---|---|---|
| | 62 | + | 12 |
| × | 61 | + | 11 |
| | $(62+11)$ | | $(12 \times 11)$ |
| | 73 | | 132 |
| | $73 \times 1/2$ | | 132 |
| | 36.5 | | 132 |
| = | $3650 + 132$ | = | 3782 |

**Ex. 9:** $54 \times 65$
**Solution:**
Base = $50 = 1/2 \times 100$

| | Number | Sign | Excess |
|---|---|---|---|
| | 54 | + | 4 |
| × | 65 | + | 15 |
| | $(54+15)$ | | $(4 \times 15)$ |
| | 69 | | 60 |
| | $69 \times 1/2$ | | 60 |
| | 34.5 | | 60 |
| = | $3450 + 60$ | = | 3510 |

**Ex. 10:** $65 \times 64$
**Solution:**
Base = $50 = 1/2 \times 100$

**Ex. 11:** $32 \times 35$
**Solution:**
Base = $30 = 3 \times 10$

| | Number | Sign | Excess | | | Number | Sign | Excess |
|---|---|---|---|---|---|---|---|---|
| | 65 | + | 15 | | | 32 | + | 2 |
| × | 64 | + | 14 | | × | 35 | + | 5 |
| | (65 + 14) | | (15 × 14) | | | (32 + 5) | | (2 × 5) |
| | 79 | | 210 | | | 37 | | 10 |
| | 79 × 1 / 2 | | 210 | | | 37 × 3 | | 10 |
| | 39.5 | | 210 | | | 111 | | 10 |
| = | 3950 + 210 | = | 4160 | | = | 1110 + 10 | = | 1120 |

Examples:

| # | Example | WB=B ÷ F | Split Result | Result/F | = |
|---|---|---|---|---|---|
| 104 | 59 × 59 | 50 = 100 ÷ 2 | 59 + 09 / 09 × 09 | 68 ÷ 2 / 81 | 3481 |
| 105 | 23 × 23 | 20 = 10 × 2 | 23 + 3 / 3 × 3 | 26 × 2 / 9 | 529 |
| 106 | 54 × 46 | 50 = 10 × 5 | 54 − 04/ − 4 × −4 | 50 × 5 / −16 | 2484 |
| 107 | 54 × 46 | 50 = 100 ÷ 2 | 54 − 04 / −04 × −04 | 50 ÷ 2 / −16 | 2484 |
| 108 | 19 × 19 | 20 = 10 × 2 | 19 − 1 / −1 × −1 | 18 × 2 / 1 | 361 |
| 109 | 23 × 21 | 20 = 10 × 2 | 23 + 1 / 3 × 1 | 24 × 2 / 3 | 483 |
| 110 | 48 × 49 | 50 = 100 ÷ 2 | 48 − 01 / −02 × −01 | 47 ÷ 2 / 02 | 2352 |
| 111 | 62 × 48 | 50 = 100 ÷ 2 | 62 − 02 / 12 × −02 | 60 ÷ 2 / −24 | 2976 |
| 112 | 249 × 248 | 250 = 1000 ÷ 4 | 248 − 001 / 001 × 002 | 247 ÷ 4 / 002 | 61752 |
| 113 | 229 × 230 | 250 = 1000 ÷ 4 | 229 − 020 / −021 × −020 | 209 ÷ 4 / 420 | 52670 |
| 114 | 19 × 499 | 500 = 1000 ÷ 2 | 19 − 001 / −481 × −001 | 18 ÷ 2 / 481 | 9481 |

## 6.5.1 Exercises for practice

**Prob. 1:** 72 × 95
**Prob. 2:** 48 × 97
**Prob. 3:** 299 × 299
**Prob. 4:** 687 × 699
**Prob. 5:** 231 × 582
**Prob. 6:** 362 × 785
**Prob. 7:** 235 × 247
**Prob. 8:** 3998 × 4998

**Prob. 9:** 635 × 502
**Prob. 10:** 389 × 516
**Prob. 11:** 532 × 472
**Prob. 12:** 523 × 522
**Prob. 13:** 889 × 9998
**Prob. 14:** 77 × 9988
**Prob. 15:** 87965 × 99998

## 6.6  Summary

In this chapter, we took the first step in our journey to unravel the speed multiplication mysteries. The method is based on understanding and manipulating with respect to a reference number, which we called the base. Using this technique, we were able to multiply rapidly without loss of accuracy. It order for us to become very good at this technique, we need to put several hours of practice. We hope we have built the required foundation to help you make progress in this journey.

# 7   Straight Multiplication

It is now time for us to go through another technique for multiplication. Speed Math is all about having an array of problem solving techniques and tools at your fingertips. With experience, you will understand which technique to use under each of the problem solving circumstances that you face.

Let us consider the product of 2 two-digit numbers. Let the numbers be *ab* and *xy*. A number *ab* is not the same as an algebraic product of two variables *a* and *b*. A number *ab* simply stands for a numerical quantity that is equal to $10a+b$. In other words, we have *a* in 10s place and *b* in units place.

1.   When we multiply a digit in Units place of a number with a digit in Units place of another number, we get a product that gives us the number of units.

2.   When we multiply a digit in Tens place of a number with a digit in Units place of another number, we get a product that gives us the number of tens. The vice versa is true as well. Digit in Units place multiplied with digit in Tens place gives us number of tens.

3.   When we multiply a digit in Tens place of a number with a digit in Tens place of another number, we get a product that gives us the number of hundreds.

What do we mean by this? If we multiply two two-digit numbers, we will have Units, Tens and Hundreds place. We can refer to the product having three parts - the Hundreds-part, the Tens-part and the Units-part.

Let us consider a simple problem. $32 \times 12 =$?

- The Hundreds-part: This comes from multiplying the digits in Tens place. $3 \times 1 = 3$.

- The Tens-part: This comes from multiplying the 3 (Tens part of first number) and 2 (Units part of the second number) and adding this to the product of 2 (Units part of first number) and 1 (Tens part of the second number). In other words, the required Tens-part of the final product is: $3 \times 2 + 2 \times 1 = 8$.

- The Units-part: This comes from the product of the Units parts of both the numbers: $2 \times 2 = 4$.

The final product therefore contains three parts: 3 hundreds, 8 tens and 2 units. This is nothing but 384.

In conventional multiplication, we would simply write this as:

$$
\begin{array}{r}
3 \ 2 \\
\times \quad 1 \ 2 \\
\hline
6 \ 4 \\
3 \ 2 \quad \\
\hline
= \ 3 \ 8 \ 4 \\
\hline
\end{array}
$$

Here the Hundreds-part, 3 is simply product of the Tens-parts of the number; Units-part of the product is simply the product of the Units parts of the numbers. Tens part comes from the sum of cross products of Units and Tens part of the two numbers.

Algebraically, we can then represent this fact as:
$$ab \times xy = (10a + b) \times (10x + y) = 100ab + 10(ay + bx) + by$$

Now is there a more intuitive way of remembering this? The answer is YES.

# 7.1 Patterns for 2 digit numbers

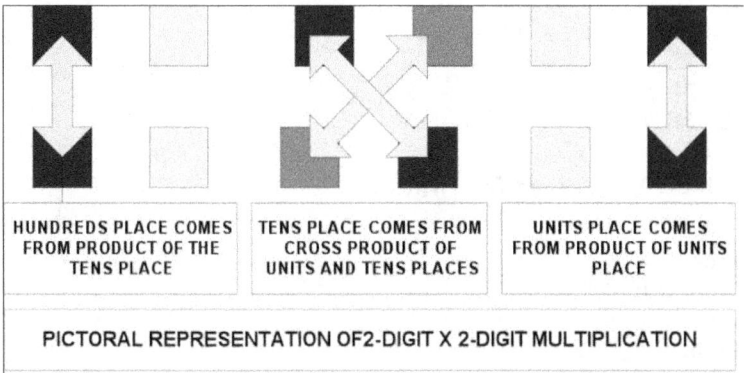

| HUNDREDS PLACE COMES FROM PRODUCT OF THE TENS PLACE | TENS PLACE COMES FROM CROSS PRODUCT OF UNITS AND TENS PLACES | UNITS PLACE COMES FROM PRODUCT OF UNITS PLACE |
|---|---|---|

PICTORAL REPRESENTATION OF 2-DIGIT X 2-DIGIT MULTIPLICATION

Clearly, the product has three parts. Therefore, we will write this as:

$$ab \times xy = \text{part-1} / \text{part-2} / \text{part-3}.$$

The slash simply indicates that there are three parts to our solution. We will drop the slashes and simply write the product in the last step.

**Ex. 1:** $12 \times 13 = 1 / 5 / 6 = 156$
**Ex. 2:** $12 \times 11 = 1 / 3 / 2 = 132$
**Ex. 3:** $16 \times 11 = 1 / 7 / 6 = 176$
**Ex. 4:** $21 \times 14 = 2 / 9 / 4 = 294$
**Ex. 5:** $23 \times 21 = 4 / 8 / 3 = 483$
**Ex. 6:** $41 \times 41 = {}_1 6 / 8 / 1 = 1681$
**Ex. 7:** $15 \times 15 = 1 / {}_1 0 / {}_2 5 = 225$
**Ex. 8:** $25 \times 25 = 4 / {}_2 0 / {}_2 5 = 625$
**Ex. 9:** $32 \times 32 = 9 / {}_1 2 / 4 = 1024$
**Ex. 10:** $35 \times 35 = 9 / {}_3 0 / {}_2 5 = 1225$
**Ex. 11:** $37 \times 33 = 9 / {}_3 0 / {}_2 1 = 1221$
**Ex. 12:** $49 \times 49 = {}_1 6 / {}_7 2 / {}_8 1 = 2401$
**Ex. 13:** $21 \times 23 = 4 / 8 / 3 = 483$
**Ex. 14:** $16 \times 13 = 1 / 9 / 18 = 208$
**Ex. 15:** $19 \times 13 = 1 / {}_1 2 / {}_2 7 = 247$
**Ex. 16:** $27 \times 14 = 2 / {}_1 5 / {}_2 8 = 378$
**Ex. 17:** $16 \times 32 = 3 / {}_2 0 / {}_1 2 = 512$
**Ex. 18:** $13 \times 21 = 2 / 7 / 3 = 273$

**Ex. 19:** $35 \times 12 = 3 \;/\; {}_1 1 \;/\; {}_1 0 = 420$
**Ex. 20:** $26 \times 14 = 2 \;/\; {}_1 4 \;/\; {}_2 4 = 364$
**Ex. 21:** $34 \times 14 = 3 \;/\; {}_1 6 \;/\; {}_1 6 = 476$
**Ex. 22:** $41 \times 51 = {}_2 0 \;/\; 9 \;/\; 1 = 2091$
**Ex. 23:** $15 \times 45 = 4 \;/\; {}_2 5 \;/\; {}_2 5 = 675$
**Ex. 24:** $24 \times 13 = 2 \;/\; {}_1 0 \;/\; {}_1 2 = 312$
**Ex. 25:** $76 \times 11 = 7 \;/\; {}_1 3 \;/\; 6 = 836$

### 7.1.1 Exercises for practice

**Prob. 1:** $24 \times 16$
**Prob. 2:** $10 \times 49$
**Prob. 3:** $99 \times 77$
**Prob. 4:** $40 \times 31$
**Prob. 5:** $61 \times 23$
**Prob. 6:** $44 \times 68$
**Prob. 7:** $12 \times 63$
**Prob. 8:** $20 \times 49$
**Prob. 9:** $15 \times 93$
**Prob. 10:** $31 \times 52$
**Prob. 11:** $30 \times 38$
**Prob. 12:** $43 \times 61$
**Prob. 13:** $43 \times 69$
**Prob. 14:** $12 \times 19$
**Prob. 15:** $33 \times 27$

## 7.2 Patterns for 3 digit numbers

Now that we have drilled the process for speed multiplication of two 2-digit numbers, we can look at patterns for 3-digit $\times$ 3-digit numbers. Let us figure the pattern out, so we can generalize the pattern in future.

3-digit $\times$ 3-digit multiplication can be represented as an expression:

$$(ax^2 + bx + c)(dx^2 + ex + f) \text{ that has 5 parts in its product.}$$

$$
\begin{aligned}
\text{Part 1} \;&=\; ad & x^4 \\
\text{Part 2} \;&=\; ae + bd & x^3 \\
\text{Part 3} \;&=\; af + be + cd & x^2 \\
\text{Part 4} \;&=\; bf + ec & x \\
\text{Part 5} \;&=\; cf & 1
\end{aligned}
$$

Let us solve the problem using this diagram. We will gather pace as we move along.

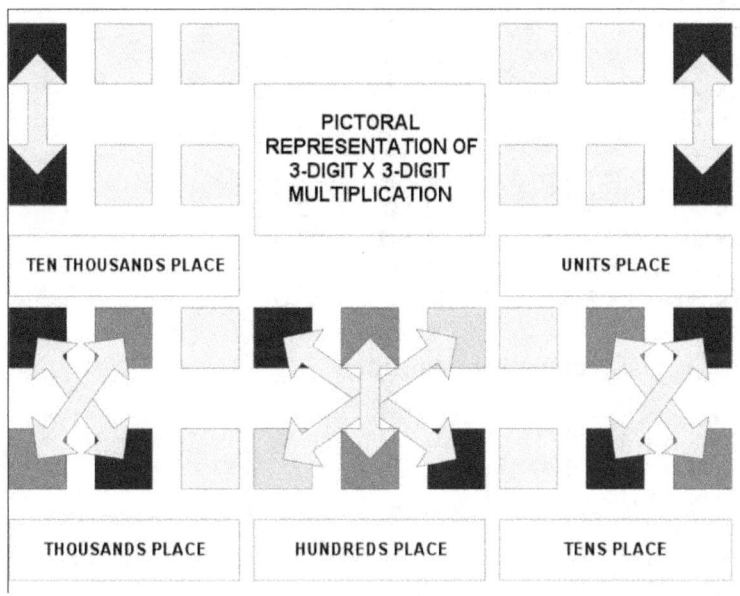

**Ex. 1:** $111 \times 111 = 1 / 2 / 3 / 2 / 1 = 12321$
**Ex. 2:** $108 \times 108 = 1 / 0 / {}_{1}6 / 0 / {}_{6}4 = 11664$
**Ex. 3:** $109 \times 111 = 1 / 1 / {}_{1}0 / 9 / 9 = 12099$
**Ex. 4:** $116 \times 114 = 1 / 2 / {}_{1}1 / {}_{1}0 / {}_{2}4 = 13224$
**Ex. 5:** $116 \times 116 = 1 / 2 / {}_{1}3 / {}_{1}2 / {}_{3}6 = 13456$
**Ex. 6:** $582 \times 231 = {}_{1}0 / {}_{3}1 / {}_{3}3 / {}_{1}4 / 2 = 134442$
**Ex. 7:** $532 \times 472 = {}_{2}0 / {}_{4}7 / {}_{3}9 / {}_{2}0 / 4 = 251104$
**Ex. 8:** $785 \times 362 = {}_{2}1 / {}_{6}6 / {}_{7}7 / {}_{4}6 / {}_{1}0 = 284170$
**Ex. 9:** $321 \times 052 = 0 / {}_{1}5 / {}_{1}6 / 9 / 2 = 16692$

**Ex. 10:** $795 \times 362 = {}_21 / {}_69 / {}_83 / {}_48 / {}_10 = 287790$
**Ex. 11:** $621 \times 547 = {}_30 / {}_34 / {}_55 / {}_18 / 7 = 339687$
**Ex. 12:** $246 \times 131 = 2 / {}_10 / {}_20 / {}_22 / 6 = 32226$
**Ex. 13:** $222 \times 146 = 2 / {}_10 / {}_22 / {}_20 / {}_12 = 32412$
**Ex. 14:** $642 \times 131 = 6 / {}_23 / {}_20 / {}_10 / 2 = 84102$
**Ex. 15:** $321 \times 213 = 6 / 7 / {}_13 / 7 / 3 = 68373$
**Ex. 16:** $889 \times 898 = {}_64 / {}_136 / {}_208 / {}_145 / {}_72 = 798322$
**Ex. 17:** $576 \times 328 = {}_15 / {}_31 / {}_72 / {}_68 / {}_48 = 188928$
**Ex. 18:** $817 \times 322 = {}_24 / {}_19 / {}_39 / {}_16 / {}_14 = 263074$
**Ex. 19:** $123 \times 121 = 1 / 4 / 8 / 8 / 3 = 14883$
**Ex. 20:** $144 \times 162 = 1 / {}_10 / {}_30 / {}_32 / 8 = 23328$
**Ex. 21:** $127 \times 354 = 3 / {}_11 / {}_35 / {}_43 / {}_28 = 44958$
**Ex. 22:** $309 \times 341 = 9 / {}_12 / {}_30 / {}_36 / 9 = 105369$
**Ex. 23:** $477 \times 121 = 4 / {}_15 / {}_25 / {}_21 / 7 = 57717$
**Ex. 24:** $147 \times 231 = 2 / {}_11 / {}_27 / {}_25 / 7 = 33957$
**Ex. 25:** $143 \times 641 = 6 / {}_28 / {}_35 / {}_16 / 3 = 91663$
**Ex. 26:** $402 \times 375 = {}_12 / {}_28 / {}_26 / {}_14 / {}_10 = 150750$
**Ex. 27:** $523 \times 423 = {}_20 / {}_18 / {}_31 / {}_12 / 9 = 221229$
**Ex. 28:** $415 \times 634 = {}_24 / {}_18 / {}_31 / {}_12 / 9 = 263110$
**Ex. 29:** $412 \times 312 = {}_12 / 7 / {}_15 / 4 / 4 = 128544$
**Ex. 30:** $423 \times 203 = 8 / 4 / {}_18 / 6 / 9 = 85869$

## 7.2.1 Exercises for practice

**Prob. 1:** $133 \times 500$
**Prob. 2:** $339 \times 764$
**Prob. 3:** $984 \times 650$
**Prob. 4:** $847 \times 559$
**Prob. 5:** $140 \times 504$
**Prob. 6:** $175 \times 964$
**Prob. 7:** $365 \times 764$
**Prob. 8:** $838 \times 643$
**Prob. 9:** $570 \times 388$
**Prob. 10:** $529 \times 476$
**Prob. 11:** $790 \times 570$
**Prob. 12:** $952 \times 356$
**Prob. 13:** $118 \times 905$
**Prob. 14:** $952 \times 653$
**Prob. 15:** $726 \times 378$

## 7.3 Patterns for 4 and 5 digit numbers

Based on our discussion in the previous section, we said, we would need to arrive at an intuitive way of identifying the parts of the product and what patterns exist. We have put the diagram for the 4-digit × 4-digit product below, and worked out some examples.

**Ex. 1:** $1123 \times 2223 = 2 \,/\, 4 \,/\, 8 \,/\, {}_15 \,/\, {}_13 \,/\, {}_12 \,/\, 9 = 2496429$
**Ex. 2:** $3150 \times 4034 = {}_12 \,/\, 4 \,/\, {}_29 \,/\, {}_15 \,/\, {}_19 \,/\, {}_20 \,/\, 0 = 12707100$
**Ex. 3:** $4000 \times 3012 = {}_12 \,/\, 0 \,/\, 4 \,/\, 8 \,/\, 0 \,/\, 0 \,/\, 0 = 12048000$
**Ex. 4:** $4233 \times 2003 = 8 \,/\, 4 \,/\, 6 \,/\, {}_18 \,/\, 6 \,/\, 9 \,/\, 9 = 8478699$

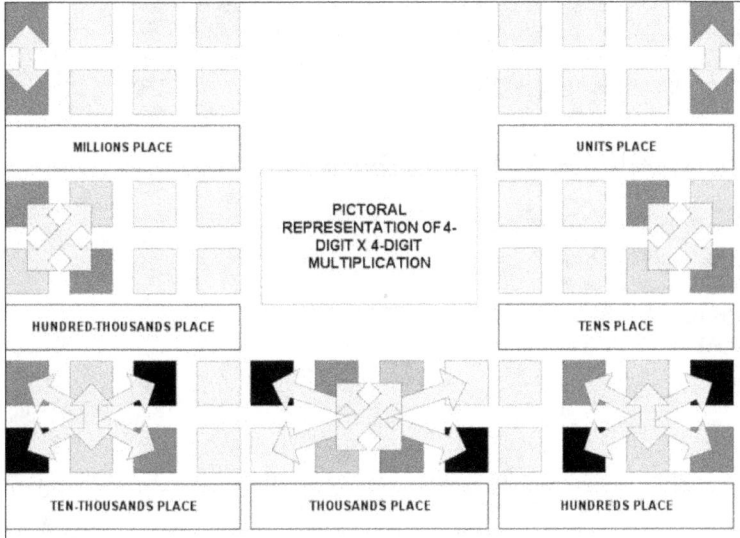

**Ex. 5:** $2710 \times 1321 = 2 \,/\, {}_13 \,/\, {}_26 \,/\, {}_19 \,/\, 9 \,/\, 1 \,/\, 0 = 3579910$
**Ex. 6:** $7100 \times 4313 = {}_28 \,/\, {}_25 \,/\, {}_10 \,/\, {}_22 \,/\, 3 \,/\, 0 \,/\, 0 = 30622300$
**Ex. 7:** $5712 \times 8320 = {}_40 \,/\, {}_71 \,/\, {}_39 \,/\, {}_33 \,/\, 8 \,/\, 4 \,/\, 0 = 47523840$
**Ex. 8:** $9438 \times 5741 = {}_45 \,/\, {}_83 \,/\, {}_79 \,/\, {}_86 \,/\, {}_72 \,/\, {}_35 \,/\, 8 = 54183558$
**Ex. 9:** $4421 \times 1234 = 4 \,/\, {}_12 \,/\, {}_22 \,/\, {}_33 \,/\, {}_24 \,/\, {}_11 \,/\, 4 = 5455514$
**Ex. 10:** $3045 \times 9874 = {}_27 \,/\, {}_24 \,/\, {}_57 \,/\, {}_89 \,/\, {}_68 \,/\, {}_51 \,/\, {}_20 = 30066330$
**Ex. 11:** $5818 \times 2505 = {}_10 \,/\, {}_41 \,/\, {}_42 \,/\, {}_46 \,/\, {}_80 \,/\, 5 \,/\, {}_40 = 14574090$
**Ex. 12:** $8178 \times 1324 = 8 \,/\, {}_25 \,/\, {}_26 \,/\, {}_63 \,/\, {}_42 \,/\, {}_44 \,/\, {}_32 = 10827672$
**Ex. 13:** $7212 \times 9021 = {}_63 \,/\, {}_18 \,/\, {}_23 \,/\, {}_29 \,/\, 4 \,/\, 5 \,/\, 2 = 65059452$

**Ex. 14:** $6403 \times 1033 = 6 / 4 / {}_18 / {}_33 / {}_12 / 9 / 9 = 6614299$
**Ex. 15:** $0364 \times 1033 = 0 / 3 / 6 / {}_13 / {}_27 / {}_30 / {}_12 = 0376012$
**Ex. 16:** $7819 \times 0212 = 0 / {}_14 / {}_23 / {}_24 / {}_35 / {}_11 / {}_18 = 1657628$
**Ex. 17:** $1176 \times 0512 = 0 / 5 / 6 / {}_38 / {}_39 / {}_20 / {}_12 = 0602112$
**Ex. 18:** $0617 \times 1321 = 0 / 6 / {}_19 / {}_22 / {}_29 / {}_15 / 7 = 0815057$
**Ex. 19:** $0053 \times 5611 = 0 / 0 / {}_25 / {}_45 / {}_23 / 8 / 3 = 0297383$
**Ex. 20:** $5216 \times 0033 = 0 / 0 / {}_15 / {}_21 / 9 / {}_21 / {}_18 = 00172128$
**Ex. 21:** $1213 \times 1121 = 1 / 3 / 5 / 9 / 7 / 7 / 3 = 1359773$
**Ex. 22:** $1402 \times 1032 = 1 / 4 / 3 / {}_16 / 8 / 6 / 4 = 1446864$
**Ex. 23:** $1200 \times 3014 = 3 / 6 / 1 / 6 / 8 / 0 / 0 = 3616800$
**Ex. 24:** $1321 \times 4012 = 4 / {}_12 / 9 / 9 / 8 / 5 / 2 = 5299852$
**Ex. 25:** $5013 \times 3102 = {}_15 / 5 / 3 / {}_20 / 3 / 2 / 6 = 15550326$
**Ex. 26:** $1112 \times 2013 = 2 / 2 / 3 / 8 / 4 / 5 / 6 = 2238456$
**Ex. 27:** $1053 \times 1041 = 1 / 0 / 9 / 4 / {}_20 / {}_17 / 3 = 1096173$
**Ex. 28:** $4302 \times 3075 = {}_12 / 9 / {}_28 / {}_47 / {}_15 / {}_14 / {}_10 = 13228650$
**Ex. 29:** $0137 \times 1280 = 0 / 1 / 5 / {}_21 / {}_38 / {}_56 / 0 = 0175360$

## 7.3.1 Exercises for practice

**Prob.** 1: $2909 \times 1924$
**Prob.** 2: $4900 \times 6894$
**Prob.** 3: $3947 \times 9277$
**Prob.** 4: $5508 \times 9754$
**Prob.** 5: $3935 \times 3426$

# 8 Multiplication using inside - outside pairs

In this section, we are going to learn a more generic multiplication technique that can handle any two numbers. In other words, we do have to worry about bases, working bases, below or above the base, deficiencies and excess. To understand this we are going to build our knowledge in a systematic manner. We will start with two two-digit numbers, and then gradually increase the number of digits as we become comfortable with this technique.

Let us spend a moment to build the basic tools and terms – which will help us understand the multiplication process better.

From the title of this chapter, it must be amply clear to all of us that we will spend some time to define and understand the concept of inside pairs and outside pairs. Understanding the outside pair and inside pair is very important, as we are going to use it frequently.

As always, we will rely on examples to understand the numerical patterns and terminologies. Experience teaches us more than theory ☺

| Problems | $46 \times 87$ | $72 \times 36$ | $97 \times 56$ | $68 \times 26$ | $81 \times 19$ |
|---|---|---|---|---|---|
| Inside Pairs | $6 \times 8$ | $2 \times 3$ | $7 \times 5$ | $8 \times 2$ | $1 \times 1$ |
| Outside Pairs | $4 \times 7$ | $7 \times 6$ | $9 \times 6$ | $6 \times 6$ | $8 \times 9$ |

A close look at the table shows the following things about the patterns:
- Inside pair is simply the product of the Units-digit of the first number and the Tens-digit of the second number
- Outside pair is the product of the Tens-digit of the first number and the Units-digit of the second number

The process is very generic. Any time you want to determine the inside and outside pairs of two two-digit numbers, we simply imagine the following arch between the two number pairs as shown below. The product of the inner arch represents the inside pair and the product of the outer arch

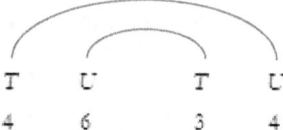

represents the outside pair. It is clear why we call it the inside pair (6 × 3 = 18) and outside pair (4 × 4 = 16).

## 8.1 Two digit multipliers

With this new approach to basic multiplication, we have to become accustomed to using a pair of digits in the multiplicand to give each figure the answer. Let us now learn to multiply any number by any number, no matter how long they are.

**Ex. 1:** 23 × 14

*Step 1*: We will add two leading zeros to 23 [in front of the multiplicand], because we have a two-digit multiplier. We will always introduce as many zeros as the digits in the multiplier. So, our game board looks like:

$$
\begin{array}{ccccccc}
& \overset{**}{} & \overset{*}{} & & \overset{*}{} & \overset{**}{} \\
0 & 0 & 2 & 3 & \times & 1 & 4
\end{array}
$$

*Step 2*: First, we will use the * mark over the digits to indicate which numbers are under consideration. The tens-digit of the multiplier and the units-digit of the multiplicand must be the outside pair, when we start the process. Therefore, the inside pair is the product of blank (right of 3 in the multiplicand) which we will treat as 0 and 1. We must pay attention to the * marks, that we have used in order to make this technique easy to understand. The outside pair is 3 × 4 = 12. Like always we will write this as 12 in the units' place of the final product.

```
            * *   *      *  * *
  0   0   2   3      ×   1   4
                  ₁2
```

*Step 3*: Now we go to the next step. The * marks move one step to the left. This is shown below. And we determine the inside pair and outside pair.

Inside pair + Outside pair = $3 \times 1 + 2 \times 4 = 11$. This sum is the current digit of the product.

```
          * *     *        *   * *
  0   0   2   3      ×   1   4
              ₁1  ₁2
```

*Step 4*: Move the *s one step to the left and repeat the process. We determine the sum of the inside and outside pairs and write them down as the current digit of the final product.

Inside pair + Outside pair = $2 \times 1 + 0 \times 4 = 2$

```
        * *   *              *  * *
  0   0   2   3      ×   1   4
      2  ₁1  ₁2
```

*Step 5*: The *s train moves one step to the left and lands over a pair of leading zeros. The sum of outside and inside pairs would be zero ($0 \times 1 + 0 \times 4$). This signifies that our process can be terminated. We now adjust for the carry forwards.

```
      * *   *                 *  * *
  0   0   2   3      ×   1   4
  0   2  ₁1  ₁2
  =   3   2   2
```

Therefore the final product is = 322.

*Step 6*: Two critical things to remember are:
1.  When the *s train starts (step 1), the inside pair product (product of the numbers with a * mark on top) is always zero.
2.  When we stop the process, both the inside and outside pairs (product of numbers with a "**" mark on top) are zero.

We can use these two flags as indicators on where to start and when to stop the process.

Examples:

**Ex. 2:** 38 × 14
**Solution:**
Add leading zeroes to the multiplicand.

| Numbers | 0 | 0 | 3 | 8 |
|---:|:---:|:---:|:---:|:---:|
| × | | | 1 | 4 |
| Inside Pair | 0 | 3×1 | 8×1 | 0×1 |
| Outside Pair | + | 0×4 | 3×4 | 8×4 |
| Partial Products | 0 | 3 | $_2 0$ | $_3 2$ |
| Answer | 0 | 5 | 3 | 2 |

**Ex. 3:** 32 × 22
**Solution:**
Add leading zeroes to the multiplicand.

| Numbers | 0 | 0 | 3 | 2 |
|---:|:---:|:---:|:---:|:---:|
| × | | | 2 | 2 |
| Inside Pair | 0 | 3×2 | 2×2 | 0×2 |
| Outside Pair | + | 0×2 | 3×2 | 2×2 |
| Partial Products | 0 | 6 | $_1 0$ | 4 |
| Answer | 0 | 7 | 0 | 4 |

**Ex. 4:** $66 \times 34$
**Solution:**
Add leading zeroes to the multiplicand.

| Numbers | 0 | 0 | 6 | 6 |
|---|---|---|---|---|
| $\times$ | | | 3 | 4 |
| Inside Pair | 0 | $6 \times 3$ | $6 \times 3$ | $0 \times 3$ |
| Outside Pair | + | $0 \times 4$ | $6 \times 4$ | $6 \times 4$ |
| Partial Products | 0 | $_1 8$ | $_4 2$ | $_2 4$ |
| Answer | 2 | 2 | 4 | 4 |

**Ex. 5:** $31 \times 15$
**Solution:**
Add leading zeroes to the multiplicand.

| Numbers | 0 | 0 | 3 | 1 |
|---|---|---|---|---|
| $\times$ | | | 1 | 5 |
| Inside Pair | 0 | $3 \times 1$ | $1 \times 1$ | $0 \times 1$ |
| Outside Pair | + | $0 \times 5$ | $3 \times 5$ | $1 \times 5$ |
| Partial Products | 0 | 3 | $_1 6$ | 5 |
| Answer | 0 | 4 | 6 | 5 |

**Ex. 6:** $34 \times 21$
**Solution:**
Add leading zeroes to the multiplicand.

| Numbers | 0 | 0 | 3 | 4 |
|---|---|---|---|---|
| $\times$ | | | 2 | 1 |
| Inside Pair | 0 | $3 \times 2$ | $4 \times 2$ | $0 \times 2$ |
| Outside Pair | + | $0 \times 1$ | $3 \times 1$ | $4 \times 1$ |
| Partial Products | 0 | 6 | $_1 1$ | 4 |
| Answer | 0 | 7 | 1 | 4 |

**Ex. 7:** 17 × 24
**Solution:**
Add leading zeroes to the multiplicand.

| | | | | |
|---|---|---|---|---|
| Numbers | 0 | 0 | 1 | 7 |
| × | | | 2 | 4 |
| Inside Pair | 0 | 1×2 | 7×2 | 0×2 |
| Outside Pair | + | 0×4 | 1×4 | 7×4 |
| Partial Products | 0 | 2 | $_1$8 | $_2$8 |
| Answer | | 4 | 0 | 8 |

**Ex. 8:** 73 × 64
**Solution:**
Add leading zeroes to the multiplicand.

| | | | | |
|---|---|---|---|---|
| Numbers | 0 | 0 | 7 | 3 |
| × | | | 6 | 4 |
| Inside Pair | 0 | 7×6 | 3×6 | 0×6 |
| Outside Pair | + | 0×4 | 7×4 | 3×4 |
| Partial Products | 0 | $_4$2 | $_4$6 | $_1$2 |
| Answer | 4 | 6 | 7 | 2 |

**Ex. 9:** 48 × 52
**Solution:**
Add leading zeroes to the multiplicand.

| | | | | |
|---|---|---|---|---|
| Numbers | 0 | 0 | 4 | 8 |
| × | | | 5 | 2 |
| Inside Pair | 0 | 4×5 | 8×5 | 0×5 |
| Outside Pair | + | 0×2 | 4×2 | 8×2 |
| Partial Products | 0 | $_2$0 | $_4$8 | $_1$6 |
| Answer | 2 | 4 | 9 | 6 |

**Ex.** 10: 312 × 14
**Solution:**
Add leading zeroes to the multiplicand.

| Numbers | 0 | 0 | 3 | 1 | 2 |
|---|---|---|---|---|---|
| × | | | | 1 | 4 |
| Inside Pair | 0 | $3 \times 1$ | $1 \times 1$ | $2 \times 1$ | $0 \times 1$ |
| Outside Pair | + | $0 \times 4$ | $3 \times 4$ | $1 \times 4$ | $2 \times 4$ |
| Partial Products | 0 | 3 | $_1 3$ | 6 | 8 |
| Answer | 0 | 4 | 3 | 6 | 8 |

**Ex.** 11: 4536 × 25
**Solution:**
Add leading zeroes to the multiplicand.

| Numbers | 0 | 0 | 4 | 5 | 3 | 6 |
|---|---|---|---|---|---|---|
| × | | | | | 2 | 5 |
| Inside Pair | 0 | $4 \times 2$ | $5 \times 2$ | $3 \times 2$ | $6 \times 2$ | $0 \times 2$ |
| Outside Pair | + | $0 \times 5$ | $4 \times 5$ | $5 \times 5$ | $3 \times 5$ | $6 \times 5$ |
| Partial Products | 0 | 8 | $_3 0$ | $_3 1$ | $_2 7$ | $_3 0$ |
| Answer | 1 | 1 | 3 | 4 | 0 | 0 |

**Ex.** 12: 4543 × 23
**Solution:**
Add leading zeroes to the multiplicand.

| Numbers | 0 | 0 | 4 | 5 | 4 | 3 |
|---|---|---|---|---|---|---|
| × | | | | | 2 | 3 |
| Inside Pair | 0 | $4 \times 2$ | $5 \times 2$ | $4 \times 2$ | $3 \times 2$ | $0 \times 2$ |
| Outside Pair | + | $0 \times 3$ | $4 \times 3$ | $5 \times 3$ | $4 \times 3$ | $3 \times 3$ |
| Partial Products | 0 | 8 | $_2 2$ | $_2 3$ | $_1 8$ | 9 |
| Answer | 1 | 0 | 4 | 4 | 8 | 9 |

**Ex.** 13: 3432 × 43
**Solution:**
Add leading zeroes to the multiplicand.

| Numbers | 0 | 0 | 3 | 4 | 3 | 2 |
|---|---|---|---|---|---|---|
| × | | | | | 4 | 3 |
| Inside Pair | 0 | $3 \times 4$ | $4 \times 4$ | $3 \times 4$ | $2 \times 4$ | $0 \times 4$ |
| Outside Pair | + | $0 \times 3$ | $3 \times 3$ | $4 \times 3$ | $3 \times 3$ | $2 \times 3$ |
| Partial Products | 0 | $_1 2$ | $_2 5$ | $_2 4$ | $_1 7$ | 6 |
| Answer | 1 | 4 | 7 | 5 | 7 | 6 |

**Ex.** 14: 1122 × 12
**Solution:**
Add leading zeroes to the multiplicand.

| Numbers | 0 | 0 | 1 | 1 | 2 | 2 |
|---|---|---|---|---|---|---|
| × | | | | | 1 | 2 |
| Inside Pair | 0 | $1 \times 1$ | $1 \times 1$ | $2 \times 1$ | $2 \times 1$ | $0 \times 1$ |
| Outside Pair | + | $0 \times 2$ | $1 \times 2$ | $1 \times 2$ | $2 \times 2$ | $2 \times 2$ |
| Partial Products | 0 | 1 | 3 | 4 | 6 | 4 |
| Answer | 0 | 1 | 3 | 4 | 6 | 4 |

**Ex.** 15: 2332 × 33
**Solution:**
Add leading zeroes to the multiplicand.

| Numbers | 0 | 0 | 2 | 3 | 3 | 2 |
|---|---|---|---|---|---|---|
| × | | | | | 3 | 3 |
| Inside Pair | 0 | $2 \times 3$ | $3 \times 3$ | $3 \times 3$ | $2 \times 3$ | $0 \times 3$ |
| Outside Pair | + | $0 \times 3$ | $2 \times 3$ | $3 \times 3$ | $3 \times 3$ | $2 \times 3$ |
| Partial Products | 0 | 6 | $_1 5$ | $_1 8$ | $_1 5$ | 6 |
| Answer | 0 | 7 | 6 | 9 | 5 | 6 |

## 8.1.1   Exercises for practice

We have drawn the grid to help you solve these problems.
**Prob.** 1: 734234 × 22
**Solution:**

| Numbers | 0 | 0 | 7 | 3 | 4 | 2 | 3 | 4 |
|---|---|---|---|---|---|---|---|---|
| × | | | | | | | 2 | 2 |
| Inside Pair | 0 | | | | | | | |
| Outside Pair | + | | | | | | | |
| Partial Products | 0 | | | | | | | |
| Answer | | | | | | | | |

**Prob.** 2: 661112 × 11
**Solution:**

| Numbers | 0 | 0 | 6 | 6 | 1 | 1 | 1 | 2 |
|---|---|---|---|---|---|---|---|---|
| × | | | | | | | 1 | 1 |
| Inside Pair | 0 | | | | | | | |
| Outside Pair | + | | | | | | | |
| Partial Products | 0 | | | | | | | |
| Answer | | | | | | | | |

**Prob.** 3: 552213 × 21
**Solution:**

| Numbers | 0 | 0 | 5 | 5 | 2 | 2 | 1 | 3 |
|---|---|---|---|---|---|---|---|---|
| × | | | | | | | 2 | 1 |
| Inside Pair | 0 | | | | | | | |
| Outside Pair | + | | | | | | | |
| Partial Products | 0 | | | | | | | |
| Answer | | | | | | | | |

**Prob.** 4: 433345 × 32
**Solution:**

| Numbers | 0 | 0 | 4 | 3 | 3 | 3 | 4 | 5 |
|---|---|---|---|---|---|---|---|---|
| × | | | | | | | 3 | 2 |
| Inside Pair | 0 | | | | | | | |
| Outside Pair | + | | | | | | | |
| Partial Products | 0 | | | | | | | |
| Answer | | | | | | | | |

**Prob.** 5: 754542 × 36
**Solution:**

| Numbers | 0 | 7 | 5 | 4 | 5 | 4 | 2 |
|---|---|---|---|---|---|---|---|
| × | | | | | | 3 | 6 |
| Inside Pair | 0 | | | | | | |
| Outside Pair | + | | | | | | |
| Partial Products | 0 | | | | | | |
| Answer | | | | | | | |

**Prob.** 6: 473254 × 43
**Solution:**

| Numbers | 0 | 0 | 4 | 7 | 3 | 2 | 5 | 4 |
|---|---|---|---|---|---|---|---|---|
| × | | | | | | | 4 | 3 |
| Inside Pair | 0 | | | | | | | |
| Outside Pair | + | | | | | | | |
| Partial Products | 0 | | | | | | | |
| Answer | | | | | | | | |

## 8.2 Three digit multipliers

We will extend the partial products that we used in the previous section. While we handled two digit multipliers, we used the inside and outside pair. The concept for three or more digits in the multipliers is the same. Let us get the concept clear, first.

Let us consider the *s train on a three digit multiplier using an example.

We now have three pairs that make up the partial products. In addition to the outside and inside pairs, we now have a middle pair, because of the three-digit multiplier.

Let us determine the product of 121 and 123.

*Step 1*: Add as many zeros in front of the multiplicand as the digits in the multiplier. Here the multiplier is 123; so we add 3 zeros in front of 121. And we lay down the game board.

```
              *** **  *     *  **  ***
   0  0  0  1  2    1        ×  1   2   3
```

*Step 2*: The partial product in the units place is 1×3 + 2×blank + 3×blank = 3. Therefore the board looks like –

```
              *** **  *     *  **  ***
   0  0  0  1  2    1        ×  1   2   3
                   3
```

*Step 3*: Now the *s train moves one step to the left.

```
            *** **  *       *  **  ***
   0  0  0  1    2  1        ×  1   2   3
                   3
```

*Step 4*: The partial product now is 2×3 (outside pair) + 1×2 (middle pair) + 1× blank (inside pair) = 8

```
           *** **  *            *  ** ***
 0  0  0  1   2  1       ×  1  2    3
              8  3
```

*Step 5*: Now the *s train moves one step to its left.

```
           *** **  *            *  ** ***
 0  0  0   1  2  1       ×  1  2    3
              8   3
```

*Step 6*: The partial product is 1×3 (outside pair) + 2×2 (middle pair) + 1×1 (inside pair) = 8

```
           *** **  *            *  ** ***
 0  0  0   1  2  1       ×  1  2    3
           8  8  3
```

*Step 7*: The *s train moves one more step to its left.

```
         *** **  *              *  ** ***
 0  0   0  1  2  1       ×  1  2    3
           8  8  3
```

*Step 8*: The partial product now is 0×3 (outside pair) + 1×2 (middle pair) + 2×1 (inside pair) = 4

```
         *** **  *              *  ** ***
 0  0   0  1  2  1       ×  1  2    3
        4  8  8  3
```

*Step 9*: Now, move the *s one more step to its left.

```
    *** **  *              *  **  ***
0   0   0  1  2  1      ×  1   2   3
           4  8  8  3
```

*Step 10*: The partial product now is 0×3 (outside pair) + 0×2 (middle pair) + 1×1 (inside pair) = 1

```
    *** **  *              *  **  ***
0   0   0  1  2  1      ×  1   2   3
        1  4  8  8  3
```

*Step 11*: Now, move the *s one more step to its left.

```
*** **  *                  *  **  ***
0   0   0  1  2  1      ×  1   2   3
        1  4  8  8  3
```

*Step 12*: Now the *s are right over the zeros. So, we can stop the process and announce the final answer.

```
*** **  *                  *  **  ***
0   0   0  1  2  1      ×  1   2   3
           1  4  8  8  3
=          1  4  8  8  3
```

Therefore the product is 14883.

We have drawn each of the mental steps explicitly. At first blush, this makes the process seem a bit complicated. Solving a few problems will ease our anxiety immensely. The elegance of the technique will be there for us to see and appreciate.

Examples:

**Ex. 1:** $1936 \times 116$
**Solution:**
Add leading zeroes to the multiplicand.

| | | | | | | | |
|---|---|---|---|---|---|---|---|
| Numbers | 0 | 0 | 0 | 1 | 9 | 3 | 6 |
| $\times$ | | | | | 1 | 1 | 6 |
| Inside Pair | 0 | $1 \times 1$ | $9 \times 1$ | $3 \times 1$ | $6 \times 1$ | $0 \times 1$ | $0 \times 1$ |
| Middle Pair | + | $0 \times 1$ | $1 \times 1$ | $9 \times 1$ | $3 \times 1$ | $6 \times 1$ | $0 \times 1$ |
| Outside Pair | + | $0 \times 6$ | $0 \times 6$ | $1 \times 6$ | $9 \times 6$ | $3 \times 6$ | $6 \times 6$ |
| Partial Products | 0 | 1 | $_1 0$ | $_1 8$ | $_6 3$ | $_2 4$ | $_3 6$ |
| Answer | 0 | 2 | 2 | 4 | 5 | 7 | 6 |

**Ex. 2:** $4183 \times 119$
**Solution:**
Add leading zeroes to the multiplicand.

| | | | | | | | |
|---|---|---|---|---|---|---|---|
| Numbers | 0 | 0 | 0 | 4 | 1 | 8 | 3 |
| $\times$ | | | | | 1 | 1 | 9 |
| Inside Pair | 0 | $4 \times 1$ | $1 \times 1$ | $8 \times 1$ | $3 \times 1$ | $0 \times 1$ | $0 \times 1$ |
| Middle Pair | + | $0 \times 1$ | $4 \times 1$ | $1 \times 1$ | $8 \times 1$ | $3 \times 1$ | $0 \times 1$ |
| Outside Pair | + | $0 \times 9$ | $0 \times 9$ | $4 \times 9$ | $1 \times 9$ | $8 \times 9$ | $3 \times 9$ |
| Partial Products | 0 | 4 | 5 | $_4 5$ | $_2 0$ | $_7 5$ | $_2 7$ |
| Answer | | 4 | 9 | 7 | 7 | 7 | 7 |

## 8.2.1 Exercises for practice

**Prob.** 1: 8922 × 105
**Solution:**

| Numbers | 0 | 0 | 0 | 8 | 9 | 2 | 2 |
|---|---|---|---|---|---|---|---|
| × | | | | | 1 | 0 | 5 |
| Inside Pair | 0 | | | | | | |
| Middle Pair | + | | | | | | |
| Outside Pair | + | | | | | | |
| Partial Products | 0 | | | | | | |
| Answer | | | | | | | |

**Prob.** 2: 2422 × 181
**Solution:**

| Numbers | 0 | 0 | 0 | 2 | 4 | 2 | 2 |
|---|---|---|---|---|---|---|---|
| × | | | | | 1 | 8 | 1 |
| Inside Pair | 0 | | | | | | |
| Middle Pair | + | | | | | | |
| Outside Pair | + | | | | | | |
| Partial Products | 0 | | | | | | |
| Answer | | | | | | | |

**Prob.** 3: 5655 × 149
**Solution:**

| Numbers | 0 | 0 | 0 | 5 | 6 | 5 | 5 |
|---|---|---|---|---|---|---|---|
| × | | | | | 1 | 4 | 9 |
| Inside Pair | 0 | | | | | | |
| Middle Pair | + | | | | | | |
| Outside Pair | + | | | | | | |
| Partial Products | 0 | | | | | | |
| Answer | | | | | | | |

Basics of Speed Mathematics

**Prob.** 4: 16337 × 132
**Solution:**

| Numbers | 0 | 0 | 0 | 1 | 6 | 3 | 3 | 7 |
|---|---|---|---|---|---|---|---|---|
| × | | | | | | 1 | 3 | 2 |
| Inside Pair | 0 | | | | | | | |
| Middle Pair | + | | | | | | | |
| Outside Pair | + | | | | | | | |
| Partial Products | 0 | | | | | | | |
| Answer | | | | | | | | |

**Prob.** 5: 36796 × 215
**Solution:**

| Numbers | 0 | 0 | 0 | 3 | 6 | 7 | 9 | 6 |
|---|---|---|---|---|---|---|---|---|
| × | | | | | | 2 | 1 | 5 |
| Inside Pair | 0 | | | | | | | |
| Middle Pair | + | | | | | | | |
| Outside Pair | + | | | | | | | |
| Partial Products | 0 | | | | | | | |
| Answer | | | | | | | | |

**Prob.** 6: 69464 × 164
**Solution:**

| Numbers | 0 | 0 | 0 | 6 | 9 | 4 | 6 | 4 |
|---|---|---|---|---|---|---|---|---|
| × | | | | | | 1 | 6 | 4 |
| Inside Pair | 0 | | | | | | | |
| Middle Pair | + | | | | | | | |
| Outside Pair | + | | | | | | | |
| Partial Products | 0 | | | | | | | |
| Answer | | | | | | | | |

**Prob.** 7: 82282 × 184
**Solution:**

| Numbers | 0 | 0 | 0 | 8 | 2 | 2 | 8 | 2 |
|---|---|---|---|---|---|---|---|---|
| × | | | | | | 1 | 8 | 4 |
| Inside Pair | 0 | | | | | | | |
| Middle Pair | + | | | | | | | |
| Outside Pair | + | | | | | | | |
| Partial Products | 0 | | | | | | | |
| Answer | | | | | | | | |

**Prob.** 8: 41961 × 105
**Solution:**

| Numbers | 0 | 0 | 0 | 4 | 1 | 9 | 6 | 1 |
|---|---|---|---|---|---|---|---|---|
| × | | | | | | 1 | 0 | 5 |
| Inside Pair | 0 | | | | | | | |
| Middle Pair | + | | | | | | | |
| Outside Pair | + | | | | | | | |
| Partial Products | 0 | | | | | | | |
| Answer | | | | | | | | |

**Prob.** 9: 77748 × 226
**Solution:**

| Numbers | 0 | 0 | 0 | 7 | 7 | 7 | 4 | 8 |
|---|---|---|---|---|---|---|---|---|
| × | | | | | | 2 | 2 | 6 |
| Inside Pair | 0 | | | | | | | |
| Middle Pair | + | | | | | | | |
| Outside Pair | + | | | | | | | |
| Partial Products | 0 | | | | | | | |
| Answer | | | | | | | | |

**Prob.** 10: 39882 × 316
**Solution:**

| Numbers | 0 | 0 | 0 | 3 | 9 | 8 | 8 | 2 |
|---|---|---|---|---|---|---|---|---|
| × | | | | | | 3 | 1 | 6 |
| Inside Pair | 0 | | | | | | | |
| Middle Pair | + | | | | | | | |
| Outside Pair | + | | | | | | | |
| Partial Products | 0 | | | | | | | |
| Answer | | | | | | | | |

**Prob.** 11: 72783 × 220
**Solution:**

| Numbers | 0 | 0 | 0 | 7 | 2 | 7 | 8 | 3 |
|---|---|---|---|---|---|---|---|---|
| × | | | | | | 2 | 2 | 0 |
| Inside Pair | 0 | | | | | | | |
| Middle Pair | + | | | | | | | |
| Outside Pair | + | | | | | | | |
| Partial Products | 0 | | | | | | | |
| Answer | | | | | | | | |

**Prob.** 12: 69495 × 131
**Solution:**

| Numbers | 0 | 0 | 0 | 6 | 9 | 4 | 9 | 5 |
|---|---|---|---|---|---|---|---|---|
| × | | | | | | 1 | 3 | 1 |
| Inside Pair | 0 | | | | | | | |
| Middle Pair | + | | | | | | | |
| Outside Pair | + | | | | | | | |
| Partial Products | 0 | | | | | | | |
| Answer | | | | | | | | |

**Prob.** 13: 54554 × 243
**Solution:**

| Numbers | 0 | 0 | 0 | 5 | 4 | 5 | 5 | 4 |
|---|---|---|---|---|---|---|---|---|
| × | | | | | | 2 | 4 | 3 |
| Inside Pair | 0 | | | | | | | |
| Middle Pair | + | | | | | | | |
| Outside Pair | + | | | | | | | |
| Partial Products | 0 | | | | | | | |
| Answer | | | | | | | | |

**Prob.** 14: 19708 × 245
**Solution:**

| Numbers | 0 | 0 | 0 | 1 | 9 | 7 | 0 | 8 |
|---|---|---|---|---|---|---|---|---|
| × | | | | | | 2 | 4 | 5 |
| Inside Pair | 0 | | | | | | | |
| Middle Pair | + | | | | | | | |
| Outside Pair | + | | | | | | | |
| Partial Products | 0 | | | | | | | |
| Answer | | | | | | | | |

**Prob.** 15: 16458 × 288
**Solution:**

| Numbers | 0 | 0 | 0 | 1 | 6 | 4 | 5 | 8 |
|---|---|---|---|---|---|---|---|---|
| × | | | | | | 2 | 8 | 8 |
| Inside Pair | 0 | | | | | | | |
| Middle Pair | + | | | | | | | |
| Outside Pair | + | | | | | | | |
| Partial Products | 0 | | | | | | | |
| Answer | | | | | | | | |

## 8.3  Summary

So far, we have seen how to multiply multiplicands of any length with a 2-digit multiplier and a 3-digit multiplier. However, when multipliers can be of any length, how do we multiply them? There is a way similar to the one we saw above. We use pair products to arrive at partial products progressively. Let us say, for example, that the multiplier has 4 digits. We will get each digit in the answer by adding up four pieces. That is there will be four pairs outside pair, two middle pairs and innermost pair. Therefore, this is the basic rule that we will follow for multiplier of any length. Let us say that we have a 5-digit multiplier, we will get each digit in the answer by adding up 5 pieces. That is: an outside pair, 3 middle pairs one below the other and an innermost pair. This can be generalized to multipliers of any length.
\

# 9 Squares and Cubes

Let us begin the discussion on squares and cubes, with a quick overview of operations. The idea of mathematical operation relates to changing of a number or a mathematical quantity from one to another. There are several examples of this. Doubling is an operation where a quantity is multiplied by 2. For example, when we double 39, we get 39 × 2, which is 78. Halving is an operation where a number is multiplied by ½. By halving 78, we get 39 back again.

Let us quickly see a few examples:

- An operation increment by 1; increases the value of a quantity by 1
- An operation decrement by 1; decreases the value of a quantity by 1

Similarly, the operation of squaring means multiply a quantity by itself. Therefore, 16-squared is 256; 19-squared is 361. The inverse operation is called the square root of a quantity. The question that we are asking is "what is the number when multiplied by it becomes the given quantity?" The square root of 256 is 16 and square root of 361 is 19.

The idea of cubes and cube roots are similar. Cube of a number is the product of number multiplied with it three times. 3-cubed is 27, 5-cubed is 125 and so on. Cube root is the inverse operation. Cube root of 125 is 5 and cube root of 27 is 3.

| Num | Num Squared | Square Root of Num Squared | Num cubed | Cube root of Num cubed |
|-----|-----|-----|-----|-----|
| 1 | 1 | 1 | 1 | 1 |
| 2 | 4 | 2 | 8 | 2 |
| 3 | 9 | 3 | 27 | 3 |
| 4 | 16 | 4 | 64 | 4 |
| 5 | 25 | 5 | 125 | 5 |
| 6 | 36 | 6 | 216 | 6 |
| 7 | 49 | 7 | 343 | 7 |
| 8 | 64 | 8 | 512 | 8 |
| 9 | 81 | 9 | 729 | 9 |

This chapter concentrates on methods of arriving at squares and cubes of numbers using a few interesting number patterns and techniques. We will first deal with squares and then move on to handling cubes.

## 9.1 Squares

Square of a number or a quantity is simply the product of the quantity multiplied with it. While the definition seems very easy to comprehend, the nature of numerical patterns that this leads us to is simply exhilarating. We will split the numbers into various classes and identify the patterns related to it.

### 9.1.1 Case 1: Numbers ending in 5

When numbers end in 5, we can write the squares of the numbers immediately.

The digit in front of the 5, in units is called first part. Therefore 35 × 35 will give an output that looks like __25. The 25 at the end of the square is the result of the 5 in the units place of the number. The digits preceding the 25 are easy to detect as well. It is simply the product of the first part with it next larger digit. The first part of 35 is 3. The next larger digit of 3 is 4. Therefore the digits preceding 25 is 3 × 4, which is 12. Therefore the square of the number is 1225.

We see that the product as two parts. The second part is 25, always. The first part is simply the product of first-part × (first-part + 1). It is that simple and straightforward. This rule is only applicable if the number ends in 5. Let us look at the table closely, where several examples of this pattern have been captured.

| Num | $1^{st}$part of Num | $1^{st}$part+1 of Num | $1^{st}$part of Product | $2^{nd}$part of Product | Square of Num |
|---|---|---|---|---|---|
| 15 | 1 | 2 | 2 | 25 | 225 |
| 25 | 2 | 3 | 6 | 25 | 625 |
| 35 | 3 | 4 | 12 | 25 | 1225 |
| 45 | 4 | 5 | 20 | 25 | 2025 |
| 55 | 5 | 6 | 30 | 25 | 3025 |
| 65 | 6 | 7 | 42 | 25 | 4225 |
| 75 | 7 | 8 | 56 | 25 | 5625 |
| 85 | 8 | 9 | 72 | 25 | 7225 |
| 95 | 9 | 10 | 90 | 25 | 9025 |
| 105 | 10 | 11 | 110 | 25 | 11025 |
| 115 | 11 | 12 | 132 | 25 | 13225 |

## 9.1.2   Case 2: 2-digit numbers beginning with 5

Here the number looks like $5n$ where $n$ is any of the single digit numbers 0-9.

The square of the number consists of two parts.
1. The second part of the square is simply the square of the digit in units place represented in two digits. 1-squared is written as 01, 2-squared as 04 and so on.
2. The first part of the square is 25 plus the number in the units place. This is the same as $25+n$.

Let us look at the table to understand the technique better.

| Num | $2^{nd}$ part of Num | $1^{st}$ part of Product | Square of Num |
|---|---|---|---|
| 51 | 01 | 25 + 1 | 2601 |
| 52 | 04 | 25 + 2 | 2704 |
| 53 | 09 | 25 + 3 | 2809 |
| 54 | 16 | 25 + 4 | 2916 |
| 55 | 25 | 25 + 5 | 3025 |
| 56 | 36 | 25 + 6 | 3136 |
| 57 | 49 | 25 + 7 | 3249 |
| 58 | 64 | 25 + 8 | 3364 |
| 59 | 81 | 25 + 9 | 3481 |

## 9.1.3   Case 3: General method of finding squares of 2-digit numbers

The general method for finding squares of 2 digit numbers can be captured in 3 easy steps. The square contains three parts; each concatenated with the other from left to right. In other words, the square appears as {first-part}{second-part}{third-part}.

Now, the first part is simply the square of the digits in the tens place, and the third part is the square of the digits in the units place. The second part is twice the product of the digits.

**Ex. 1:** $32^2 = \{3^2\}\ \{2 \times 3 \times 2\}\ \{2^2\} = 9\ _12\ 4 = 1024$
**Ex. 2:** $49^2 = \{4^2\}\ \{2 \times 4 \times 9\}\ \{9^2\} = 16\ _72\ _81 = 2401$
**Ex. 3:** $63^2 = \{6^2\}\ \{2 \times 6 \times 3\}\ \{3^2\} = 36\ _36\ 9 = 3969$
**Ex. 4:** $68^2 = \{6^2\}\ (2 \times 6 \times 8\}\ \{8^2\} = 36\ _96\ _64 = 4624$
**Ex. 5:** $75^2 = \{7^2\}\ \{2 \times 7 \times 5\}\ \{5^2\} = 49\ _70\ _25 = 5625$
**Ex. 6:** $72^2 = \{7^2\}\ \{2 \times 7 \times 2\}\ \{2^2\} = 49\ _28\ 4 = 5184$
**Ex. 7:** $84^2 = \{8^2\}\ \{2 \times 8 \times 4\}\ \{4^2\} = 64\ _64\ 16 = 7056$
**Ex. 8:** $88^2 = \{8^2\}\ \{2 \times 8 \times 8\}\ \{8^2\}\ 64\ _{128}\ _64 = 7744$
**Ex. 9:** $91^2 = \{9^2\}\ \{2 \times 9 \times 1\}\ \{1^2\} = 81\ _18\ 1 = 8281$
**Ex. 10:** $94^2 = \{9^2\}\ \{2 \times 9 \times 4\}\ \{4^2\} = 81\ _72\ _16 = 8836$

## 9.1.4 Exercises for practice

**Prob. 1:** $37^2$
**Prob. 2:** $26^2$
**Prob. 3:** $66^2$
**Prob. 4:** $44^2$
**Prob. 5:** $97^2$
**Prob. 6:** $77^2$
**Prob. 7:** $47^2$
**Prob. 8:** $76^2$
**Prob. 9:** $69^2$
**Prob. 10:** $84^2$

## 9.1.5 Case 3: Squaring 3-digit numbers

This follows the same principle like the 3 digit multiplication. The representation of the technique in English may make it sound a bit complicated. However, with a little practice, you can master this technique in no time. Let us dive into the technique right away.

Let us consider a three digit number $abc$. This is not to be confused with algebraic term $abc$ where it stands for product of $a$, $b$ and $c$. The number $abc$ simply means that $a$ is in the hundreds place, $b$ in the tens place and $c$ is in the units place. The numerical value of number $abc$ is $100a + 10b + c$. Hope this clarifies things.

$abc^2$ contains 5 parts in the product which can be represented as:

$$abc^2 = \{a^2\}\{2\times a\times b\}\{b^2 + 2\times a\times c\}\{2\times b\times c\}\{c^2\}$$

Therefore, we can determine $621^2$ using the above thumb rule.

**Ex. 1:** $621^2 = \{6^2\}\ \{2\times6\times2\}\ (2^2 + 2\times6\times1)\ \{2\times2\times1\}\ \{1^2\}$
$= 36\ _24\ _16\ 4\ 1 = 385641$
**Ex. 2:** $789^2 = \{7^2\}\ \{2\times7\times8\}\ \{8^2 + 2\times7\times9\}\ \{2\times8\times9\}\ \{9^2\}$
$= 49\ _{11}2\ _{19}0\ _{14}4\ _81 = 622521$
**Ex. 3:** $324^2 = \{3^2\}\ \{2\times3\times2\}\ \{2^2 + 2\times3\times4\}\ \{2\times2\times4\}\ \{4^2\}$
$= 9\ _12\ _28\ _16\ _16 = 104976$
**Ex. 4:** $556^2 = \{5^2\}\ \{2\times5\times5\}\ (5^2 + 2\times5\times6)\ \{2\times5\times6\}\ \{6^2\}$
$= 25\ _50\ _85\ _60\ _36 = 309136$

**Ex. 5:** $682^2 = \{6^2\}\ \{2\times6\times8\}\ \{8^2 + 2\times6\times2\}\ \{2\times8\times2\}\ \{2^2\}$
$= 36\ _96\ _88\ _32\ 4 = 465124$

**Ex. 6:** $499^2 = \{4^2\}\ \{2\times4\times9\}\ \{9^2 + 2\times4\times9\}\ \{2\times9\times9\}\ \{9^2\}$
$= 16\ _72\ _{15}3\ _{16}2\ _81 = 249001$

**Ex. 7:** $876^2 = \{8^2\}\ \{2\times8\times7\}\ \{7^2 + 2\times8\times6\}\ \{2\times7\times6\}\ \{6^2\}$
$= 64\ _{11}2\ _{14}5\ _84\ 36 = 767376$

**Ex. 8:** $936^2 = \{9^2\}\ \{2\times9\times3\}\ \{3^2 + 2\times9\times6\}\ \{2\times3\times6\}\ \{6^2\}$
$= 81\ _54\ _{11}7\ _36\ 36 = 876096$

**Ex. 9:** $289^2 = \{2^2\}\ \{2\times2\times8\}\ \{8^2 + 2\times2\times9\}\ \{2\times8\times9\}\ \{9^2\}$
$= 4\ _32\ _{10}0\ _{14}4\ _81 = 83521$

**Ex. 10:** $316^2 = \{3^2\}\ \{2\times3\times1\}\ \{1^2 + 2\times3\times6\}\ \{2\times1\times6\}\ \{6^2\}$
$= 9\ 6\ _37\ _12\ 36 = 99856$

## 9.1.6 Exercises for practice

**Prob. 1:** $448^2$
**Prob. 2:** $331^2$
**Prob. 3:** $222^2$
**Prob. 4:** $165^2$
**Prob. 5:** $832^2$
**Prob. 6:** $714^2$
**Prob. 7:** $256^2$
**Prob. 8:** $192^2$
**Prob. 9:** $801^2$
**Prob. 10:** $123^2$

## 9.2 Cubing

Cubing of 2-digit numbers is relatively straightforward. Let us consider a two digit number '$ab$'; which is numerically equal to "$10a + b$".

In order to determine $ab^3$, we simply write the 4 terms:

$$ab^3$$

| $a^3$ | $a^2 \times b$ | $a \times b^2$ | $b^3$ |
|---|---|---|---|
| + | $2 \times a^2 \times b$ | $2 \times a \times b^2$ | |
| = | | | |

Add the terms and you will get the cube of the three digit number in four parts.

**Ex. 1:** $11^3$

$11^3$

|   | 1 | 1 | 1 | 1 |
|---|---|---|---|---|
| + |   | 2 | 2 |   |
|   | 1 | 3 | 3 | 1 |
| = | 1 | 3 | 3 | 1 |

**Ex. 2:** $12^3$

$12^3$

|   | 1 | 2 | 4 | 8 |
|---|---|---|---|---|
| + |   | 4 | 8 |   |
|   | 1 | 6 | $_1$2 | 8 |
| = | 1 | 7 | 2 | 8 |

**Ex. 3:** $13^3$

$13^3$

|   | 1 | 3 | 9 | 27 |
|---|---|---|---|----|
| + |   | 6 | 18 |   |
|   | 1 | 9 | $_2$7 | $_2$7 |
| = | 2 | 1 | 9 | 7 |

**Ex. 4:** $14^3$

$14^3$

|   | 1 | 4 | 16 | 64 |
|---|---|---|----|----|
| + |   | 8 | 32 |   |
|   | 1 | $_1$2 | $_4$8 | $_6$4 |
| = | 2 | 7 | 4 | 4 |

**Ex. 5:** $15^3$

$15^3$

|   | 1 | 5 | 25 | 125 |
|---|---|---|----|-----|
| + |   | 10 | 50 |   |
|   | 1 | $_1$5 | $_7$5 | $_{12}$5 |
| = | 3 | 3 | 7 | 5 |

**Ex. 6:** $16^3$

$16^3$

|   | 1 | 6 | 36 | 216 |
|---|---|---|----|-----|
| + |   | 12 | 72 |   |
|   | 1 | $_1$8 | $_{10}$8 | $_{21}$6 |
| = | 4 | 0 | 9 | 6 |

**Ex. 7:** $17^3$

$17^3$

|   | 1 | 7 | 49 | 343 |
|---|---|---|----|-----|
| + |   | 14 | 98 |   |
|   | 1 | $_2$1 | $_{14}$7 | $_{34}$3 |
| = | 4 | 9 | 1 | 3 |

**Ex. 8:** $19^3$

$19^3$

|   | 1 | 9 | 81 | 729 |
|---|---|---|----|-----|
| + |   | 18 | 162 |   |
|   | 1 | $_2$7 | $_{24}$3 | $_{72}$9 |
| = | 6 | 8 | 5 | 9 |

**Ex. 9:** $21^3$

$21^3$

|   |   | 8 | 4 | 2 | 1 |
|---|---|---|---|---|---|
| + |   |   | 8 | 4 |   |
|   |   | 8 | $_1$2 | 6 | 1 |
| = |   | 9 | 2 | 6 | 1 |

**Ex. 10:** $22^3$

$22^3$

|   |   | 8 | 8 | 8 | 8 |
|---|---|---|---|---|---|
| + |   |   | 16 | 16 |   |
|   |   | 8 | $_2$4 | $_2$4 | 8 |
| = | 1 | 0 | 6 | 4 | 8 |

## 9.2.1 Exercises for practice

**Prob. 1:** $23^3$

$23^3$

+

=

**Prob. 2:** $24^3$

$24^3$

+

=

**Prob. 3:** $25^3$

$25^3$

+

=

**Prob. 4:** $32^3$

$32^3$

+

=

**Prob. 5:** $33^3$

$33^3$

+

=

Prob. 6: 413

$41^3$

+

=

**Prob.** 7: $42^3$

$42^3$ _____

+ _____

= _____


**Prob.** 8: $51^3$

$51^3$ _____

+ _____

= _____


**Prob.** 9: $52^3$

$52^3$ _____

+ _____

= _____


**Prob.** 10: $97^3$

$97^3$ _____

+ _____

= _____


## 9.3 General Technique for Squaring

We will now look into a general technique for determining squares of arbitrarily large numbers. But such a process always starts with a single digit number and proceeds systematically to handle numbers with increasing number of digits.

We will define a function called $D$ [for duplex], which works in the following way.

$D(a) = a^2$, where $a$ is a single digit number

$D(ab) = 2 \times a \times b$, where $ab$ is a two digit number

$D(abc) = 2 \times a \times c + b^2$, where $abc$ is a three digit number

$D(abcd) = 2 \times a \times d + 2 \times b \times c$, where $abcd$ is a four digit number

$D(abcde) = 2 \times a \times e + 2 \times b \times d + c^2$, where $abcde$ is a five digit number

Square of a single digit number is simply the duplex of that number or $D$(single-digit number). Therefore: $2^2 = D(2) = 2 \times 2 = 4$.

The same function can be applied for the other single digit numbers as well.

The interesting application of the duplex function to the problem of squaring actually starts from the 2-digit numbers onwards.

$ab^2$ consists of three parts: $ab^2 = D(a) / D(ab) / D(b)$.

Let us look at the table below. We recommend that you solve these examples instead of reading through the table. This will provide the much-required practice.

| $ab^2$ | $D(a)$ | $D(ab)$ | $D(b)$ | Int. Answer | $=$ |
|---|---|---|---|---|---|
| $87^2$ | $8 \times 8 = 64$ | $2 \times 8 \times 7 = 112$ | $7 \times 7 = 49$ | $64 /_{11} 2 /_4 9$ | 7569 |
| $31^2$ | $3 \times 3 = 9$ | $2 \times 3 \times 1 = 6$ | $1 \times 1 = 1$ | $9 / 6 / 1$ | 961 |
| $64^2$ | $6 \times 6 = 36$ | $2 \times 6 \times 4 = 48$ | $4 \times 4 = 16$ | $36 /_4 8 /_1 6$ | 4096 |
| $15^2$ | $1 \times 1 = 1$ | $2 \times 1 \times 5 = 10$ | $5 \times 5 = 25$ | $1 /_1 0 /_2 5$ | 225 |
| $76^2$ | $7 \times 7 = 49$ | $2 \times 7 \times 6 = 84$ | $6 \times 6 = 36$ | $49 /_8 4 /_3 6$ | 5776 |
| $69^2$ | $6 \times 6 = 36$ | $2 \times 6 \times 9 = 108$ | $9 \times 9 = 81$ | $36 /_{10} 8 /_8 1$ | 4761 |
| $49^2$ | $4 \times 4 = 16$ | $2 \times 4 \times 9 = 72$ | $9 \times 9 = 81$ | $16 /_7 2 /_8 1$ | 2401 |
| $30^2$ | $3 \times 3 = 9$ | $2 \times 3 \times 0 = 0$ | $0 \times 0 = 0$ | $9 / 0 / 0$ | 900 |
| $33^2$ | $3 \times 3 = 9$ | $2 \times 3 \times 3 = 18$ | $3 \times 3 = 9$ | $9 /_1 8 / 9$ | 1089 |
| $92^2$ | $9 \times 9 = 81$ | $2 \times 9 \times 2 = 36$ | $2 \times 2 = 4$ | $81 /_3 6 / 4$ | 8464 |

Similarly, the squares of a three digit numbers have 5 parts in them. Let $abc$ be a three digit number. Then:

$abc^2 = D(a) / D(ab) / D(abc) / D(bc) / D(c)$

Adjusting for carry forwards, we get the final answer. Let us look at a few examples of how to go about this process. Like before, we will capture the steps in a table.

| $abc^2$ | $D(a)$ | $D(ab)$ | $D(abc)$ | $D(bc)$ | $D(c)$ | Int. Answer | = |
|---|---|---|---|---|---|---|---|
| $201^2$ | 4 | 0 | 4 | 0 | 1 | $4/0/4/0/1$ | 40401 |
| $874^2$ | 64 | 112 | 113 | 56 | 16 | $64/_{11}2/_{11}3/_{5}6/_{1}6$ | 763876 |
| $317^2$ | 9 | 6 | 43 | 14 | 49 | $9/6/_{4}3/_{1}4/_{4}9$ | 100489 |
| $641^2$ | 36 | 48 | 28 | 8 | 1 | $36/_{4}8/_{2}8/8/1$ | 410881 |
| $153^2$ | 1 | 10 | 31 | 30 | 9 | $1/_{1}0/_{3}1/_{3}0/9$ | 23409 |
| $767^2$ | 49 | 84 | 134 | 84 | 49 | $49/_{8}4/_{13}4/_{8}4/_{4}9$ | 588289 |
| $691^2$ | 36 | 108 | 93 | 18 | 1 | $36/_{10}8/_{9}3/_{1}8/1$ | 477481 |
| $497^2$ | 16 | 72 | 137 | 126 | 49 | $16/_{7}2/_{13}7/_{12}6/_{4}9$ | 247009 |
| $306^2$ | 9 | 0 | 36 | 0 | 36 | $9/0/_{3}6/0/_{3}6$ | 93636 |
| $337^2$ | 9 | 18 | 51 | 42 | 49 | $9/_{1}8/_{5}1/_{4}2/_{4}9$ | 113569 |

Moving on to the problem of handling squares of 4-digit numbers, we can see that 4 digit numbers have 7 parts in the intermediate answer, each from a corresponding duplex function.

$$abcd^2 = D(a)\,/\,D(ab)\,/\,D(abc)\,/\,D(abcd)\,/\,D(bcd)\,/\,D(cd)\,/\,D(d)$$

| $abcd^2$ | $D(a)$ | $D(ab)$ | $D(abc)$ | $D(abcd)$ | $D(bcd)$ | $D(cd)$ | $D(d)$ | Intermediate Answer | = |
|---|---|---|---|---|---|---|---|---|---|
| $2791^2$ | 4 | 28 | 85 | 130 | 95 | 18 | 1 | $4/_{2}8/_{8}5/_{13}0/_{9}5/_{1}8/1$ | 7789681 |
| $4105^2$ | 16 | 8 | 1 | 40 | 10 | 0 | 25 | $16/8/1/_{1}0/_{1}0/0/_{2}5$ | 16851025 |
| $4202^2$ | 16 | 16 | 4 | 16 | 8 | 0 | 4 | $16/_{1}6/4/_{1}6/8/0/4$ | 17656804 |
| $5073^2$ | 25 | 0 | 70 | 30 | 49 | 42 | 9 | $25/0/_{7}0/_{3}0/_{4}9/_{4}2/9$ | 25735329 |
| $3203^2$ | 9 | 12 | 4 | 18 | 12 | 0 | 9 | $9/_{1}2/4/_{1}8/_{1}2/0/9$ | 10259209 |
| $2751^2$ | 4 | 28 | 69 | 74 | 39 | 10 | 1 | $4/_{2}8/_{6}9/_{7}4/_{3}9/_{1}0/1$ | 7568001 |
| $3536^2$ | 9 | 30 | 43 | 66 | 69 | 36 | 36 | $9/_{3}0/_{4}3/_{6}6/_{6}9/_{3}6/_{3}6$ | 12503296 |
| $5084^2$ | 25 | 0 | 80 | 40 | 64 | 64 | 16 | $25/0/_{8}0/_{4}0/_{6}4/_{6}4/_{1}6$ | 25847056 |
| $3793^2$ | 9 | 42 | 103 | 144 | 123 | 54 | 9 | $9/_{4}2/_{10}3/_{14}4/_{12}3/_{5}4/9$ | 14386849 |
| $2345^2$ | 4 | 12 | 25 | 44 | 46 | 40 | 25 | $4/_{1}2/_{2}5/_{4}4/_{6}0/_{4}0/_{2}5$ | 5499025 |

Similarly, the square of a 5-digit number can be represented by the duplex functions as shown below:

$$abcde^2 = D(a)\,/\,D(ab)\,/\,D(abc)\,/\,D(abcd)\,/\,D(abcde)$$
$$/\,D(bcde)\,/\,D(cde)\,/\,D(de)\,/\,D(e)$$

| $abcde^2$ | $a$ | $ab$ | $abc$ | $abcd$ | $abcde$ | $bcde$ | $cde$ | $de$ | $e$ | Answer |
|---|---|---|---|---|---|---|---|---|---|---|
| $12691^2$ | 1 | 4 | ${}_1 6$ | ${}_4 2$ | ${}_7 4$ | ${}_{11} 2$ | ${}_9 3$ | ${}_1 8$ | 1 | 161061481 |
| $43475^2$ | 16 | ${}_2 4$ | ${}_4 1$ | ${}_8 0$ | ${}_9 8$ | ${}_8 6$ | ${}_8 9$ | ${}_7 0$ | ${}_2 5$ | 1890075625 |
| $42903^2$ | 16 | ${}_1 6$ | ${}_7 6$ | ${}_3 6$ | ${}_{10} 5$ | ${}_1 2$ | ${}_5 4$ | 0 | 9 | 1840667409 |
| $16266^2$ | 1 | ${}_1 2$ | ${}_4 0$ | ${}_3 6$ | ${}_8 8$ | ${}_9 6$ | ${}_6 0$ | ${}_7 2$ | ${}_3 6$ | 264582756 |
| $34088^2$ | 9 | ${}_2 4$ | ${}_1 6$ | ${}_4 8$ | ${}_{11} 2$ | ${}_6 4$ | ${}_6 4$ | ${}_{12} 8$ | ${}_6 4$ | 1161991744 |
| $14224^2$ | 1 | 8 | ${}_2 0$ | ${}_2 0$ | ${}_2 8$ | ${}_4 0$ | ${}_2 0$ | ${}_1 6$ | ${}_1 6$ | 202322176 |
| $30552^2$ | 9 | 0 | ${}_3 0$ | ${}_3 0$ | ${}_3 7$ | ${}_5 0$ | ${}_4 5$ | ${}_2 0$ | 4 | 933424704 |
| $14827^2$ | 1 | 8 | ${}_3 2$ | ${}_6 8$ | ${}_9 4$ | ${}_8 8$ | ${}_{11} 6$ | ${}_2 8$ | ${}_4 9$ | 219839929 |
| $51326^2$ | 25 | ${}_1 0$ | ${}_3 1$ | ${}_2 6$ | ${}_7 3$ | ${}_2 4$ | ${}_4 0$ | ${}_2 4$ | ${}_3 6$ | 2634358276 |
| $12546^2$ | 1 | 4 | ${}_1 4$ | ${}_2 8$ | ${}_5 3$ | ${}_6 4$ | ${}_7 6$ | ${}_4 8$ | ${}_3 6$ | 157402116 |

## 9.3.1 Exercises for practice

| $abcde^2$ | $a$ | $ab$ | $abc$ | $abcd$ | $abcde$ | $bcde$ | $cde$ | $de$ | $e$ | Answer |
|---|---|---|---|---|---|---|---|---|---|---|
| $19917^2$ | | | | | | | | | | |
| $20284^2$ | | | | | | | | | | |
| $10577^2$ | | | | | | | | | | |
| $35129^2$ | | | | | | | | | | |
| $26276^2$ | | | | | | | | | | |
| $41623^2$ | | | | | | | | | | |
| $15527^2$ | | | | | | | | | | |
| $43528^2$ | | | | | | | | | | |
| $23625^2$ | | | | | | | | | | |

**Prob. 11:** If we had a 7-digit number *abcdef*, identify the duplex functions for $abcdef^2$

# 10 One subtle change to improve your accuracy

We will commence our discussion on division with a little note on how our accuracy of division can be improved dramatically by making a few changes to the way we are taught to divide numbers at school.

For the process to be repeatable, easy and less error prone, it must be systematic and clean. It must be easy to identify the errors in case they creep in during the problem solving process.

Let us explore the process by solving a few problems.

**Ex. 1:** 27483624 ÷ 62

We are going to divide 27483624 by 62. The setup we use is most similar to our conventional method.

62)      27483624        (Quotient here

The process of division simply requires us to check how many times the divisor can be repeatedly subtracted from a partial dividend. So, we build a column of figures under the divisor by repeatedly adding 62 to the number above. This gives us the basic tables of 62 that we will use to look up during the division process. Here is an easy way to get this done quickly.

| Times | Divisor | Step | Order of building the table quickly |
|-------|---------|------|-------------------------------------|
| 1 | 62 | 0 | This is the divisor |
| 2 | 124 | 1 | Double the divisor |
| 3 | 186 | 9 | Add divisor to Step 1 |
| 4 | 248 | 2 | Double result in Step 1 |
| 5 | 310 | 5 | Halve the result of Step 4 |
| 6 | 372 | 6 | Add divisor to Step 5 |
| 7 | 434 | 7 | Add divisor to Step 6 |
| 8 | 496 | 3 | Double the result in Step 2 |
| 9 | 558 | 8 | Add divisor to Step 3 |
| 10 | 620 | 4 | Add 0 at the end of the divisor |

During the process of division the numbers in the Times column will produce the partial quotients.

Now we can proceed to the process of actual division.

The procedure is now similar to the division method we are taught at school. The thought process has been captured in the form of steps.

*Step 1*: The divisor (62) is less than 2 and 27. Therefore we consider the first three digits of the dividend. This is 274.

*Step 2*: What is the largest number in divisor table that is less than 274? The answer is 248. Therefore, the partial quotient is 4. We write 4 in the quotient field and subtract 248 from 274, giving us a current remainder of 26.

*Step 3*: Now we bring down the next digit of the dividend – 8. We have 268. We will ask the same question as before. What is the largest number in the divisor table that is less than 268? The answer is 248. Therefore, this step produces a partial quotient of 4 as well. Subtracting 248 from 268, we get 20.

*Step 4*: Now we bring down the next digit in the dividend. We have 203. We repeatedly look up the divisor table to ensure that we are picking up the largest number that is less than current remainder.

*Next Steps*: By repeating the aforementioned steps, we get the final quotient of 443284 with a remainder of 16.

The detailed working of the solution is shown below.

| Times | Divisor | Q: | | 4 | 4 | 3 | 2 | 8 | 4 |
|---|---|---|---|---|---|---|---|---|---|
| 1 | 62 | | 2 | 7 | 4 | 8 | 3 | 6 | 2 | 4 |
| 2 | 124 | − | 2 | 4 | 8 | | | | |
| 3 | 186 | | | 2 | 6 | 8 | | | |
| 4 | 248 | | − | 2 | 4 | 8 | | | |
| 5 | 310 | | | 2 | 0 | 3 | | | |
| 6 | 372 | | | − | 1 | 8 | 6 | | |
| 7 | 434 | | | 1 | 7 | 6 | | | |
| 8 | 496 | | | − | 1 | 2 | 4 | | |
| 9 | 558 | | | | 5 | 2 | 2 | | |
| 10 | 620 | | | − | 4 | 9 | 6 | | |
| | | | | | 2 | 6 | 4 | | |
| | | | | − | 2 | 4 | 8 | | |
| | | | | | | 1 | 6 | | |

The basic question we ask repeatedly is – "what is the largest number in the divisor column which is less than or equal to the current remainder or partial dividend in question?" The corresponding figure in the "Times" column gives us the partial quotient.

What if we picked the wrong number from the divisor table? Here is an illustration from the previous example.

Let us consider the very first step to illustrate the possible errors and how we can ensure that we are on the right track. We started with considering 274 as the partial dividend and determined that the largest number in the divisor column that is less than or equal to this number is 248 – which gives us a quotient of 4.

What happened if we picked up 186 instead? When we subtract 186 from 274, we would be left with 88. 88 is greater than the divisor. This means that we can still subtract the divisor from the remainder at least one more time. 88 − 62 = 24. Now 24 is lesser than 62; which gives us the correct current quotient of 4. This gives us the first check in the division process. *The partial remainder throughout the process must always be less than the divisor.*

Let us assume that we missed this check and proceeded to the next step. We bring the next digit of the dividend down, which is 8 and we now have 888 as the current dividend. There largest number in our divisor table that is less than 888 in 620, which gives us a current quotient of 10. This gives us the second check in the division process. *The current quotient will always be a single digit number. If we ever get a two-digit coefficient, we must realize that we have made a mistake* and check our steps thoroughly.

*The purpose of the last entry of 620 in the divisor table is exactly this. If we ever have to use this number in our division process, we can stop and be sure that there is an error in computation.*

The subtle introduction of the divisor table makes the process of division – easy, accurate and mechanical. Error detection and identification is simple. Let us solve a few examples to get better at this. With practice, you will see that the few additional seconds that you use up in order to build a divisor table is well worth the effort.

Let us summarize the steps and capture our key observations.

Form a divisor table by adding the divisor repeatedly up to ten times.

We try to keep in mind three rules to get this division done:
- *Rule* 1: Identify the largest number in the divisor table that is lesser than or equal to the current dividend.
- *Rule* 2: Subtract this number from the current dividend. The result must be lesser than the divisor. If it is greater than or equal to the divisor, then we have certainly made a mistake. Time for us to go back to rule 1. ☺
- *Rule* 3: The corresponding entry in the Times column gives us the partial quotient.

More examples:

**Ex. 2:** 6876 ÷ 8

| Times | Divisor | Q: | | 8 | 5 | 9 |
|---|---|---|---|---|---|---|
| 1 | 8 | 6 | 8 | 7 | 6 | |
| 2 | 16 | − | 6 | 4 | | |
| 3 | 24 | | | 4 | 7 | |
| 4 | 32 | − | | 4 | 0 | |
| 5 | 40 | | | | 7 | 6 |
| 6 | 48 | | − | | 7 | 2 |
| 7 | 56 | | | | | 4 |
| 8 | 64 | | | | | |
| 9 | 72 | | | | | |
| 10 | 80 | | | | | |

Hence the quotient = 859, remainder = 4.

**Ex. 3:** 897899 ÷ 87

| Times | Divisor | Q: | | 1 | 0 | 3 | 2 | 0 |
|---|---|---|---|---|---|---|---|---|
| 1 | 87 | 8 | 9 | 7 | 8 | 9 | 9 | |
| 2 | 174 | − | 8 | 7 | | | | |
| 3 | 261 | | | 2 | 7 | 8 | | |
| 4 | 348 | − | | 2 | 6 | 1 | | |
| 5 | 435 | | | | 1 | 7 | 9 | |
| 6 | 522 | | | − | 1 | 7 | 4 | |
| 7 | 609 | | | | | | 5 | 9 |
| 8 | 696 | | | | | | | |
| 9 | 783 | | | | | | | |
| 10 | 870 | | | | | | | |

Hence the quotient = 10320, remainder = 59.

**Ex. 4:** 66754 ÷ 31

| Times | Divisor | Q: | | 2 | 1 | 5 | 3 |
|---|---|---|---|---|---|---|---|
| 1 | 31 | | 6 | 6 | 7 | 5 | 4 |
| 2 | 62 | − | 6 | 2 | | | |
| 3 | 93 | | | 4 | 7 | | |
| 4 | 124 | | − | 3 | 1 | | |
| 5 | 155 | | | 1 | 6 | 5 | |
| 6 | 186 | | − | 1 | 5 | 5 | |
| 7 | 217 | | | | 1 | 0 | 4 |
| 8 | 248 | | | − | | 9 | 3 |
| 9 | 279 | | | | | 1 | 1 |
| 10 | 310 | | | | | | |

Hence the quotient = 2153, remainder = 11.

**Ex. 5:** 545653 ÷ 55

| Times | Divisor | Q: | | | 9 | 9 | 2 | 0 |
|---|---|---|---|---|---|---|---|---|
| 1 | 55 | | 5 | 4 | 5 | 6 | 5 | 3 |
| 2 | 110 | − | 4 | 9 | 5 | | | |
| 3 | 165 | | | 5 | 0 | 6 | | |
| 4 | 220 | | − | 4 | 9 | 5 | | |
| 5 | 275 | | | | 1 | 1 | 5 | |
| 6 | 330 | | | − | 1 | 1 | 0 | |
| 7 | 385 | | | | | | 5 | 3 |
| 8 | 440 | | | | | | | |
| 9 | 495 | | | | | | | |
| 10 | 550 | | | | | | | |

Hence the quotient = 9920, remainder = 53

**Ex. 6:** 778668 ÷ 93

| Times | Divisor | Q: | | | 8 | 3 | 7 | 2 |
|---|---|---|---|---|---|---|---|---|
| 1 | 93 | | 7 | 7 | 8 | 6 | 6 | 8 |
| 2 | 186 | − | 7 | 4 | 4 | | | |
| 3 | 279 | | | 3 | 4 | 6 | | |
| 4 | 372 | | − | 2 | 7 | 9 | | |
| 5 | 465 | | | | 6 | 7 | 6 | |
| 6 | 558 | | | − | 6 | 5 | 1 | |
| 7 | 651 | | | | | 2 | 5 | 8 |
| 8 | 744 | | | | − | 1 | 8 | 6 |
| 9 | 837 | | | | | | 7 | 2 |
| 10 | 930 | | | | | | | |

Hence the quotient = 8372, remainder = 72.

**Ex. 7:** 8788 ÷ 5

| Times | Divisor | Q: | 1 | 7 | 5 | 7 |
|---|---|---|---|---|---|---|
| 1 | 5 | | 8 | 7 | 8 | 8 |
| 2 | 10 | − | 5 | | | |
| 3 | 15 | | 3 | 7 | | |
| 4 | 20 | − | 3 | 5 | | |
| 5 | 25 | | | 2 | 8 | |
| 6 | 30 | | − | 2 | 5 | |
| 7 | 35 | | | | 3 | 8 |
| 8 | 40 | | | − | 3 | 5 |
| 9 | 45 | | | | | 3 |
| 10 | 50 | | | | | |

Hence the quotient = 1757, remainder = 3.

**Ex.** 8: 93421 ÷ 69

| Times | Divisor | Q: | | 1 | 3 | 5 | 3 |
|---|---|---|---|---|---|---|---|
| 1 | 69 | | 9 | 3 | 4 | 2 | 1 |
| 2 | 138 | − | 6 | 9 | | | |
| 3 | 207 | | 2 | 4 | 4 | | |
| 4 | 276 | − | 2 | 0 | 7 | | |
| 5 | 345 | | | 3 | 7 | 2 | |
| 6 | 414 | − | | 3 | 4 | 5 | |
| 7 | 483 | | | | 2 | 7 | 1 |
| 8 | 552 | − | | | 2 | 0 | 7 |
| 9 | 621 | | | | | 6 | 4 |
| 10 | 690 | | | | | | |

Hence the quotient = 1353, remainder = 64

**Ex.** 9: 39541 ÷ 84

| Times | Divisor | Q: | | 4 | 7 | 0 |
|---|---|---|---|---|---|---|
| 1 | 84 | | 3 | 9 | 5 | 4 | 1 |
| 2 | 168 | − | 3 | 3 | 6 | | |
| 3 | 252 | | 5 | 9 | 4 | | |
| 4 | 336 | − | 5 | 8 | 8 | | |
| 5 | 420 | | | 6 | 1 | | |
| 6 | 504 | | | | | | |
| 7 | 588 | | | | | | |
| 8 | 672 | | | | | | |
| 9 | 756 | | | | | | |
| 10 | 840 | | | | | | |

Hence the quotient = 470, remainder = 61

**Ex.** 10: 710309 ÷ 75

| Times | Divisor | Q: | | | | 9 | 4 | 7 | 0 |
|---|---|---|---|---|---|---|---|---|---|
| 1 | 75 | | 7 | 1 | 0 | 3 | 0 | 9 | |
| 2 | 150 | − | 6 | 7 | 5 | | | | |
| 3 | 225 | | | 3 | 5 | 3 | | | |
| 4 | 300 | | − | 3 | 0 | 0 | | | |
| 5 | 375 | | | | 5 | 3 | 0 | | |
| 6 | 450 | | | − | 5 | 2 | 5 | | |
| 7 | 525 | | | | | 5 | 9 | | |
| 8 | 600 | | | | | | | | |
| 9 | 675 | | | | | | | | |
| 10 | 750 | | | | | | | | |

Hence the quotient = 9470, remainder = 59

**Ex.** 11: 729716 ÷ 6

| Times | Divisor | Q: | 1 | 2 | 1 | 6 | 1 | 9 |
|---|---|---|---|---|---|---|---|---|
| 1 | 6 | | 7 | 2 | 9 | 7 | 1 | 6 |
| 2 | 12 | − | 6 | | | | | |
| 3 | 18 | | 1 | 2 | | | | |
| 4 | 24 | − | 1 | 2 | | | | |
| 5 | 30 | | | 0 | 9 | | | |
| 6 | 36 | | | − | 6 | | | |
| 7 | 42 | | | | 3 | 7 | | |
| 8 | 48 | | | − | 3 | 6 | | |
| 9 | 54 | | | | | 1 | 1 | |
| 10 | 60 | | | | − | | 6 | |
| | | | | | | | 5 | 6 |
| | | | | | | − | 5 | 4 |
| | | | | | | | | 2 |

Hence the quotient = 121619, remainder = 2

179

# 11 Basic Patterns in Division

We commence our journey to explore the patterns in division by focusing on division by 9 using several examples. This process will give us several clues about the nature of the division process.

While dividing a 2-digit number by 9, we will draw a vertical line in between the tens and units place. We will call this the R-line. We will get the remainder to the right of the R-line and coefficient to the left of it. This R-line will highlight an important property of division.

We will try to solve 12 ÷ 9. We will write the divisor to the left. The dividend is written to the right. Now that we have drawn the game board, we will go to the next step of creating a thumb rule for division by 9.

- We draw the R-line, in between the units' and tens' place of the dividend.
- We simply move the digit in the 10s place of the dividend under the units place.

We add the digits on either side of the R-line. The sum to the left of the R-line is the quotient and the sum to the right of the R-line is the remainder.

**Ex. 1:** 12 ÷ 9

| 9) | 1 | 2 | Draw the R-line |
|---|---|---|---|
| | | 1 | Move the 10s place to units place |
| | 1 | 3 | Add the numbers |
| | $Q = 1$ | $R = 3$ | |

The Q stands for quotient and R stands for remainder. Let us use this newfound understanding and solve several examples. This will help internalize the procedure.

**Ex. 2:** $21 \div 9$

| 9) | 2 | 1 | Draw the R-line |
|---|---|---|---|
| | | 2 | Move the 10s place to units place |
| | 2 | 3 | Add the numbers |
| | $Q = 2$ | $R = 3$ | |

**Ex. 3:** $33 \div 9$

| 9) | 3 | 3 | Draw the R-line |
|---|---|---|---|
| | | 3 | Move the 10s place to units place |
| | 3 | 6 | Add the numbers |
| | $Q = 3$ | $R = 6$ | |

**Ex. 4:** $40 \div 9$

| 9) | 4 | 0 | Draw the R-line |
|---|---|---|---|
| | | 4 | Move the 10s place to units place |
| | 4 | 4 | Add the numbers |
| | $Q = 4$ | $R = 4$ | |

**Ex. 5:** $52 \div 9$

| 9) | 5 | 2 | Draw the R-line |
|---|---|---|---|
| | | 5 | Move the 10s place to units place |
| | 5 | 7 | Add the numbers |
| | $Q = 5$ | $R = 7$ | |

**Ex. 6:** $61 \div 9$

| 9) | 6 | 1 | Draw the R-line |
|---|---|---|---|
| | | 6 | Move the 10s place to units place |
| | 6 | 7 | Add the numbers |
| | $Q = 6$ | $R = 7$ | |

**Ex.** 7: 70 ÷ 9

```
9)        7 |  0          Draw the R-line
            |  7    Move the 10s place to units place
          7 |  7          Add the numbers
       ─────────────
       Q = 7 | R = 7
```

**Ex.** 8: 80 ÷ 9

```
9)        8 |  0          Draw the R-line
            |  8    Move the 10s place to units place
          8 |  8          Add the numbers
       ─────────────
       Q = 8 | R = 8
```

The examples so far have demonstrated the applicability of this technique to two digit numbers. Let us now extend the technique to 3 digit dividends.

Here we repeatedly take the first part of the dividend, the part to the left of the R-line; till we arrive at solving a two-digit division by 9. We simply add the columns on either side of the R-line as always.

If we have to handle 103÷9, we put 10 to the left of the R-line and 3 to the right of the R-line. We then solve 10÷9; which results in 1 to the left of the R-line and 1 to the right of the R-line. This is shown in the following example.

**Ex.** 9: 103 ÷ 9

```
9)       10 |  3      Draw the R-line
          1 |  1       Solve 10 ÷ 9
       ─────────────
         11 |  4    Add the numbers
       ─────────────
      Q = 11 | R = 4
```

**Ex.** 10: 113 ÷ 9

```
9)       11 |  3      Draw the R-line
          1 |  2       Solve 11 ÷ 9
       ─────────────
         12 |  5    Add the numbers
       ─────────────
      Q = 12 | R = 5
```

The 11÷9 step is captured in short hand. You can do this in detail as well.

| 9) | 11 | 3 | Draw the R-line |
|---|---|---|---|
| | 1 | 1 | In 11 ÷ 9:10s digit at left and 1s at right of R-line |
| | | 1 | Move the 10s digit to the right of R-line |
| | 12 | 5 | Add the numbers |
| | $Q = 12$ | $R = 5$ | |

Both these are valid approaches to handling the situation. With practice you should simply start looking at a two digit number $ab \div 9$ and see that it produces a quotient of $a$ and a remainder of $a+b$. Therefore, we can simply put the quotient to the left of the R-line and remainder to the right of the R-line.

**Ex. 11:** $124 \div 9$

| 9) | 12 | 4 | Draw the R-line |
|---|---|---|---|
| | 1 | 3 | 12 ÷ 9 gives Q=1; R=3 |
| | 13 | 7 | Add the numbers |
| | $Q = 13$ | $R = 7$ | |

**Ex. 12:** $211 \div 9$

| 9) | 21 | 1 | Draw the R-line |
|---|---|---|---|
| | 2 | 3 | 21 ÷ 9 gives Q=2; R=3 |
| | 23 | 4 | Add the numbers |
| | $Q = 23$ | $R = 4$ | |

**Ex. 13:** $160 \div 9$

| 9) | 16 | 0 | Draw the R-line |
|---|---|---|---|
| | 1 | 7 | 16 ÷ 9 gives Q=1; R=7 |
| | 17 | 7 | Add the numbers |
| | $Q = 17$ | $R = 7$ | |

**Ex. 14:** 311 ÷ 9

| 9) | 31 | 1 | Draw the R-line |
|---|---|---|---|
| | 3 | 4 | 31 ÷ 9 gives Q=3; R=4 |
| | 34 | 5 | Add the numbers |
| | Q = 34 | R = 5 | |

We will now handle a four-digit dividend. The divisor continues to be 9.

**Ex. 15:** 1203 ÷ 9

| 9) | 120 | 3 | Draw the R-line |
|---|---|---|---|
| | 12 | 0 | 120 ÷ 9 |
| | 1 | 3 | 12 ÷ 9 gives Q=1; R=3 |
| | 133 | 6 | Add the numbers |
| | Q = 133 | R = 6 | |

**Ex. 16:** 1230 ÷ 9

| 9) | 123 | 0 | Draw the R-line |
|---|---|---|---|
| | 12 | 3 | 123 ÷ 9 |
| | 1 | 3 | 12 ÷ 9 gives Q=1; R=3 |
| | 136 | 6 | Add the numbers |
| | Q = 136 | R = 6 | |

**Ex. 17:** 120021 ÷ 9

| 9) | 12002 | 1 | Draw the R-line |
|---|---|---|---|
| | 1200 | 2 | 12002 ÷ 9 |
| | 120 | 0 | 1200 ÷ 9 |
| | 12 | 0 | 120 ÷ 9 |
| | 1 | 3 | 12 ÷ 9 gives Q=1; R=3 |
| | 13335 | 6 | Add the numbers |
| | Q = 13335 | R = 6 | |

The procedure for handling division by 9 is simple and straightforward. In fact, we are now able to handle very large dividends. Let us spend a moment to make a few interesting observations.

- There is no multiplication and subtraction; unlike conventional division that we are used to.
- We simply draw an R-line and the quotient and remainders seems to gather on either side of this line.
- We do need to perform a few simple addition operations.

Yes, at this point, we know that this works when divisor is 9. How we wish that this be extended for other divisors as well?

Let us consider a tricky situation. $18 \div 9$ highlights this case.

**Ex. 18:** $18 \div 9$

| 9) | 1 | 8 | Draw the R-line |
|---|---|---|---|
| | | 1 | Move the 10s place to units place |
| | 1 | 9 | Add the numbers |
| | $Q = 1$ | $R = 9$ | Since remainder greater than the divisor, |
| | $Q = 2$ | $R = 0$ | subtract divisor from remainder and increment quotient |

'Remainder must always be less than the divisor. If it is greater than or equal to the divisor, we have to subtract the divisor from the remainder and increment the quotient. We must repeat this process until remainder is less than the divisor.

Therefore in the previous example, we must take away 9 from remainder and increment the quotient. This gives us Q=2; R=0. This is an essential last step in our process.

This is the only place in our division procedure where we have used a subtraction operation! Can you beat that?

**Ex. 19:** $225 \div 9$

| 9) | 22 | 5 | Draw the R-line |
|---|---|---|---|
| | 2 | 4 | $22 \div 9$ gives Q=2; R=4 |
| | 24 | 9 | Add the numbers |
| | $Q = 24$ | $R = 9$ | R=9, hence do R-9 and increment Q |
| | $Q = 25$ | $R = 0$ | |

**Ex.** 20: $136 \div 9$

| 9) | 13 | 6 | Draw the R-line |
|---|---|---|---|
| | 1 | 4 | $13 \div 9$ gives Q=1; R=4 |
| | 14 | 10 | Add the numbers |
| | $Q = 14$ | $R = 10$ | R>9, hence do R-9 and increment Q |
| | $Q = 15$ | $R = 1$ | |

Let us now summarize our findings:

1. The process seems to work well when the divisor is 9
2. There is no subtraction in most cases; a simple addition is all that is required.
3. Sum of the digits in the number is the remainder
4. Quotient is simply a progressive sum of the first parts of successive dividends.
5. We seem to be dealing with small number; therefore the probability of making a mistake is reduced.
6. Is this procedure applicable to divisors other than 9? This is the key question.

In the subsequent sections of this topic, we will find answers to this question. In order to answer this question, we will solve a few examples to understand the patterns.

Now, we will start considering divisors other than 9. We will refer to the number to the left of the R-line as the first part and the number to the right of the R-line as the second part. This will help us define things with ease.

**Ex.** 21: $23 \div 8$

| 8) | 2 | 3 | Draw the R-line |
|---|---|---|---|
| | | 4 | First part × 10s complement of divisor: 2 × 2=4 |
| | 2 | 7 | Add the numbers |
| | $Q = 2$ | $R = 7$ | |

**Ex. 22:** $12 \div 7$

| 7) | 1 | 2 | Draw the R-line |
|---|---|---|---|
| | | 3 | First part × 10s complement of divisor: 1 × 3=3 |
| | 1 | 5 | Add the numbers |
| | $Q = 1$ | $R = 5$ | |

**Ex. 23:** $11 \div 6$

| 6) | 1 | 1 | Draw the R-line |
|---|---|---|---|
| | | 4 | First part × 10s complement of divisor: 1 × 4=4 |
| | 1 | 5 | Add the numbers |
| | $Q = 1$ | $R = 5$ | |

Interestingly, this works when divisor is 9. 10s complement of 9 is 1. Therefore, first part multiplied by 10s complement of 9 is the same as first part itself.

Thus we have found an elegant way of handling simple division problems in this section. Let us work through the same examples and look at the steps in a different light. This alternative viewpoint will help us consolidate our understanding and summarize the underlying principle.

## 11.1  New way of looking at division process

**Ex. 1:** $12345 \div 9$

| 9) | 1 | 2 | 3 | 4 | 5 |
|---|---|---|---|---|---|
| 1 | | 1 | 3 | 6 | 10 |
| | | 1 | 3 | 6 | ₁0 | 15 |
| = | 1 | 3 | 7 | 0 | 15 |
| = | 1 | 3 | 7 | 1 | 6 |

*Step 1:* The 10s complement of 9 is 1. This is captured right below the divisor. The number of digits in the complement is 1.

Therefore, the R-line is drawn in between the units and the tens place of the dividend. In other words, our remainder will have 1 digit as well. This is not surprising. We know that the final remainder must be less than the divisor. A number less than 9 will have 1 digit only!

*Step 2*: Bring down the first digit of the dividend as is. This is the first digit of the partial quotient.

*Step 3*: Multiply the partial quotient from the previous with the 10s complement of the divisor. We will refer to this as the partial product. Write partial product under the next digit of the dividend. Add the numbers in the column, i.e. the next digit of the dividend and the partial product. This gives us the next digit of the partial quotient.

*Step 4*: Repeat Step 3 until to run out of digits in the dividend.

*Step 5*: The quotient is 1370 and remainder is 15. 15 is greater than 9. The remainder cannot be greater than the divisor. Therefore, we increase the quotient by 1 and decrease the remainder by 9. Therefore, the quotient is 1371 and remainder is 6.

We recommend that you solve the problem using the method described in the previous section and compare the steps. You will see that the previous method is the same as this one. Let us consider one more example to ensure that we have understood the nuts and bolts of this technique.

**Ex.** 2: 10117 ÷ 9

| 9) | 1 | 0 | 1 | 1 | 7 |
|----|---|---|---|---|---|
| 1 |   | 1 | 1 | 2 | 3 |
|    | 1 | 1 | 2 | 3 | 10 |
| = | 1 | 1 | 2 | 4 | 1 |

We draw the R-Line, because we know that we will get a one-digit remainder. The solution process commences with bringing own the first digit of the dividend, as is. The partial quotient we get is 1. Now we create the partial product by multiplying the partial quotient with the complement; and write it down under the next digit of the dividend. Add the column of numbers to get next partial quotient. We continue this process until we run out of digits in the dividend. The entire solution is displayed above. We get a remainder of 10, which is greater than the divisor. Therefore increment the quotient by 1 and reduce the remainder by 9. This leads us to the final answer: Quotient = 1124, Remainder = 1.

**Ex. 3:** 26785 ÷ 9

```
9)  2   6   7   8 | 5
    1       2   8  15 | 23
            2   8  15  23 | 28
    =   2   9   7   3 | 28
    =   2   9   7   6 | 1
```

Quotient = 2976; Remainder = 1

Let us now consider the case of a single digit divisor other than 9. This should set the stage for further discussions on the topic.

**Ex. 4:** 136 ÷ 8

```
8)  1   3 | 6
    2       2 | 10
            1   5 | 16
    =   1   5 | 16
    =   1   7 | 0
```

Quotient = 17; Remainder = 0

The process is the same as before. We just remember to multiply the partial quotient with 2; because the 10s complement of 8 is 2. R-line ensures that the remainder is a single digit number. The process works for divisors other than 9.

So, let us solve a few problems to ensure that our understanding is correct. And of course, practice makes us perfect.

**Ex. 5:** 12314 ÷ 6

```
6)   1   2    3     1 | 4
4        4   24   108 | 436
         1    6   27  109 | 440
   =  1   9    7    9 | 440


6)                4    4 | 0
4                     16 | 20
                 4   20 | 20
                 6    0 | 20
                 6    3 | 2
```

Quotient = 1979, Remainder = 440

Now we will do $440 \div 6$

```
6)   4    4 | 0
4        16 | 80
     4   20 | 80
 =   6    0 | 80
```

Quotient = 60, Remainder = 80
Now $80 \div 6 = 13$, Remainder=2

Therefore, Quotient = 1979 + 60 + 13 = 2052; Remainder = 2

Clearly, if the remainder is very large, we can repeat the steps again, with the remainder as the new dividend and complete the division process. This technique is very generic.

## 11.2 Summary

Let us summarize the highlights of this method and also understand what the key issues are.
1. The process works well for all single digit divisors.
2. There is a simple multiplication step [complement × partial quotient).
3. There is no subtraction step at all.

4.  However, as the divisor moves away from the base, the complement increases. This results larger products. And therefore larger remainders.
5.  When the divisor is close to the base, this works very well. Its quick and accurate.
6.  If the remainder is a large number, we simply treat the remainder as the new dividend and repeat the process.

Now comes the big question. How do we handle larger divisors? We will take a short detour into basic algebra and see how we can divide algebraic expressions. This will give us some insights into handling divisors with 2 or more digits.

Here is the clue. An algebraic expression $x^2 + 2x + 1$ is the same as 121 when $x$ is equal to 10. Every polynomial expression in x is a decimal number at $x = 10$. We might have to handle carry forwards when the coefficients are greater than 9, but this important linkage will be exploited to explain the procedure for handling divisors with 2 or more digits.

Here is a fun filled exercise. While solving each of the problems in the next section, solve the same problem assuming that you are dealing with a decimal number with $x = 10$. The analogy throws more light on the problem solving approach that we are about to take.

# 12 Extension to the Speed Division

## 12.1 Algebraic Expressions

In this section, we will begin with algebraic expressions and how we would divide one expression with another. If you have no idea of what algebra is at this point in your academic journey, please do not lose heart. Just go one to the later half of this section and solve problems assuming that they somehow work like magic. Pretty much like the way we all learnt to ride a bicycle without any idea of Newton's laws of motion ☺. You will eventually understand the technique after a brief course in elementary algebra.

There are a few observations on the relationship between a polynomial in $x$ and a number. A polynomial in $x$ represents a number when $x = 10$. For example, $x + 1$ is the same as 11 when $x = 10$. Similarly, $x - 1$ is the same as 9 when $x = 10$. Similarly, $x^2 + 2x + 1$ is the same as 121 at $x = 10$. Getting a feel for this equivalence between the algebraic and arithmetic worlds will help us to gain insights and expertise in problem solving. Therefore, we can view an algebraic expression; a polynomial in $x$ is simply a number in base $x$, much like the decimal number system which is in base 10.

Therefore the following statements are true.
- 10s complement of 9 is 1.
- 10s complement of 8 is 2.
- 10s complement of 6 is 4.

Similarly,
1. $x$s complement of $x - 1$ is 1.
2. $x$s complement of $x - 3$ is 3.
3. $x$s complement of $x - n$ is $n$.
4. $x$s complement of $x + n$ is $-n$.

Armed with this understanding of the linkages between the two worlds, we can now take another step towards understanding speed division. Let is start with a problem.

**Ex. 1:** $(12x^2 - 8x - 32) \div (x - 2)$

$x$-complement of $(x - 2)$ is 2.

The table below captures the essence of this method. We will follow this up with detailed explanations.

|  |  | $x^2$ | $x$ | 1 |
|---|---|---|---|---|
| Divisor | $x - 2$ | 12 | $-8$ | $-32$ |
| Complement | 2 |  | $12 \times 2 = 24$ | $16 \times 2 = 32$ |
| Quotient |  | $12x^2 \div x = 12x$ | 16 | $R = 0$ |

*Step 1*: Divide the first term of the dividend ($12x^2$) by the first term of the divisor ($x$). We get $12x$. We capture this as the first term of the quotient.

*Step 2:* We multiply the complement of the divisor (2) with the coefficient of the first term in the quotient (12). This product (24) is added with the coefficient of the second term of the dividend (-8). We get 16. This is the second term of the quotient. Since the first term in the quotient was $x$; the next term is the constant. Therefore, we can now move on to determining the remainder.

*Step 3:* We multiply of the complement of the divisor (2) with the second term (16). This product (32) is added to the third term of the dividend (-32). We get remainder, which this case is $32 - 32 = 0$.

Therefore, $Q = 12x + 16$; $R = 0$.

Let us continue to solve a few more problems and the technique will become clearer as we make progress.

**Ex. 2:** $(7x^2 + 5x + 3) \div (x - 1)$

The solution has been captured in the table:

|  |  | $x^2$ | $x$ | 1 |
|---|---|---|---|---|
| Divisor | $x - 1$ | 7 | 5 | 3 |
| Complement | 1 |  | $1 \times 7 = 7$ | $1 \times 12 = 12$ |
| Quotient |  | $7 \div 1 = 7$ | $5 + 7 = 12$ | $R = 15$ |

*Step 1*: Divide the first term of the dividend ($7x^2$) by the first term of the divisor ($x$). We get $7x$. We capture this as the first term of the quotient. This is the same as $7 \div 1 = 7$

*Step 2*: We multiply the complement of the divisor (1) with the coefficient of the first term in the quotient (7). This product (7) is added with the coefficient of the second term of the dividend (5). We get 12. This is the second term of the quotient. Since the first term in the quotient was $x$; the next term is the constant. Therefore, we can now move on to determining the remainder.

*Step 3:* We multiply of the complement of the divisor (1) with the second term (12). This product (12) is added to the third term of the dividend (3). We get remainder, which this case is $3 + 12 = 15$.

Therefore, $Q = 7x + 12; R = 15$.

Now that you get the general drift of the process, we will capture the solution and let us check and see if this makes sense.

**Ex. 3:** $(x^3 - x^2 + 7x + 3) \div (x - 3)$

|  | | $x^3$ | $x^2$ | $x$ | 1 |
|---|---|---|---|---|---|
| Divisor | $x - 3$ | 1 | $-1$ | 7 | 3 |
| Complement | 3 | | $3 \times 1 = 3$ | $3 \times 2 = 6$ | $3 \times 13 = 39$ |
| Quotient | | $1 \div 1 = 1$ | $-1 + 3 = 2$ | $7 + 6 = 13$ | $R = 42$ |

Therefore, $Q = x^2 + 2x + 13; R = 42$.

Let us summarize our observations of this technique.
1. Some patterns need to be noticed and understood. The first term of the quotient is $x^2$. This means that subsequent term be $x$ and next term will be a constant. The final term will be the remainder.
2. While capturing the coefficients of the dividend, it is important that we capture the coefficients of every decreasing power of $x$. This means that a dividend $x^3 + x + 1$ must be captured as $x^3 + 0x^2 + x + 1$.
3. This procedure works when the coefficient of the first term in the divisor is 1. We will gain practice making this assumption. We will extend the technique for the case when the coefficient of the first term of the divisor is not equal to 1.

**Ex. 4:** $(7x^2 + 5x + 3) \div (x + 1)$

|            |        | $x^2$       | $x$               | 1                    |
|------------|--------|-------------|-------------------|----------------------|
| Divisor    | $x + 1$ | 7           | 5                 | 3                    |
| Complement | $-1$   |             | $-1 \times 7 = -7$ | $-1 \times -2 = 2$   |
| Quotient   |        | $7 \div 1 = 7$ | $5 - 7 = -2$      | $R = 5$              |

Therefore, Q = $7x$ – 2; R = 5

**Ex. 5:** $(x^3 + 7x^2 + 6x + 5) \div (x - 2)$

|            |        | $x^3$       | $x^2$          | $x$               | 1                |
|------------|--------|-------------|----------------|-------------------|------------------|
| Divisor    | $x - 2$ | 1           | 7              | 6                 | 5                |
| Complement | 2      |             | $2 \times 1 = 2$ | $2 \times 9 = 18$ | $2 \times 24 = 48$ |
| Quotient   |        | $1 \div 1 = 1$ | $7 + 2 = 9$    | $18 + 6 = 24$     | $R = 53$         |

Therefore, Q = $x^2$ + 9$x$ + 24; R = 53

**Ex. 6:** $(x^3 - x^2 + 7x + 3) \div (x + 3)$

|            |        | $x^3$       | $x^2$             | $x$                | 1                  |
|------------|--------|-------------|-------------------|--------------------|--------------------|
| Divisor    | $x + 3$ | 1           | $-1$              | 7                  | 3                  |
| Complement | $-3$   |             | $-3 \times 1 = -3$ | $-3 \times -4 = 12$ | $-3 \times 19 = -57$ |
| Quotient   |        | $1 \div 1 = 1$ | $-1 - 3 = -4$     | $7 + 12 = 19$      | $R = -54$          |

Therefore, Q = $x^2$ – 4$x$ + 19; R = –54

**Ex. 7:** $(x^3 + 3x^2 + 10x - 7) \div (x - 5)$

|            |        | $x^3$       | $x^2$          | $x$               | 1                   |
|------------|--------|-------------|----------------|-------------------|---------------------|
| Divisor    | $x - 5$ | 1           | 3              | 10                | $-7$                |
| Complement | 5      |             | $5 \times 1 = 5$ | $5 \times 8 = 40$ | $5 \times 50 = 250$ |
| Quotient   |        | $1 \div 1 = 1$ | $3 + 5 = 8$    | $10 + 40 = 50$    | $R = 243$           |

Therefore, Q = $x^2$ +8$x$ + 50; R = 243

**Ex. 8:** $(x^4 - 3x^3 + 7x^2 + 5x + 7) \div (x - 4)$

|  |  | $x^4$ | $x^3$ | $x^2$ | $x$ | 1 |
|---|---|---|---|---|---|---|
| Divisor | $x-4$ | 1 | $-3$ | 7 | 5 | 7 |
| Complement | 4 |  | $4 \times 1 = 4$ | $4 \times 1 = 4$ | $4 \times 11 = 44$ | $4 \times 49 = 196$ |
| Quotient |  | $1 \div 1 = 1$ | $-3 + 4 = 1$ | $4 + 7 = 11$ | $5 + 44 = 49$ | $R = 203$ |

Therefore, $Q = x^3 + x^2 + 11x + 49$; $R = 203$

We will now extend this technique to the case where we have more terms. The process is similar to what we have used so far. Let us remind ourselves how we determine the complements. The complement of $x^2 - x + 1$ is $x + 1$. We simply drop the first term and reverse the signs of the other terms of the expression. The resulting term or expression gives us the complement.

**Ex. 9:** $(x^4 - x^3 + x^2 + 3x + 5) \div (x^2 - x - 1)$

|  |  | $x^4$ | $x^3$ | $x^2$ | $x$ | 1 |
|---|---|---|---|---|---|---|
| Divisor | $x^2 - x - 1$ | 1 | $-1$ | 1 | 3 | 5 |
| Complement | $1 + 1$ |  | 1 | 1 |  |  |
|  |  |  |  | 0 | 0 |  |
|  |  |  |  |  | 2 | 2 |
| Quotient |  | $1 \div 1 = 1$ | 0 | 2 | $R = 5$ | 7 |

Therefore, $Q = x^2 + 2$; $R = 5x + 7$

The complement $1 + 1$ simply stands for $x + 1$. Therefore, when we multiply the complement with the coefficient of the current partial quotient, we have to multiply each of the terms of the complement and write them down under appropriate "place values". For example, $1 + 1$ multiplied by 1 gives us $1 + 1$. Therefore, we write 1 under each of $x^3$ and $x^2$ columns. We then add $-1$ and 1 in the $x^3$ column. This gives us a current quotient term of 0. The next step is to multiply $1 + 1$ with 0. We get $0 + 0$. These go under $x^2$ and $x$ columns. The sum of the terms in the $x^2$ column is 2. This becomes our coefficient of the current term in our quotient. Finally, we multiply $1 + 1$ with 2 and write down $2 + 2$ under $x$ and the constant columns. We thus compute the remainder.

We can summarize our observations.

1. The process is pretty much the same as before. The coefficient of the first term of the divisor must be 1.

2. The complement of the divisor is determined by using a simple thumb rule. We drop the first term and reverse the signs of the other terms. This gives us the complement of the divisor.

3. Rest of the process is similar to the one used before. The power of $x$ in the first term of the coefficient is simply determined by dividing the first term of the dividend with the first term of the divisor. Each subsequent term in the quotient has diminishing powers of $x$.

4. The term after the constant is the beginning of the remainder expression. Here is a thumb rule for this as well. If the highest power of the dividend is 4, then the dividend has 5 terms in all. We will have 4 terms with diminishing powers of $x$ and a constant. If the highest power of $x$ in the divisor is 2, the divisor will similarly have 3 terms. This means that the complement of the divisor will have one term less than the divisor.

5. The remainder simply has as many terms as the complement of the divisor. It is therefore possible for us to predict where the quotient ends and where the remainder begins using this thumb rule. This means that we can draw an R-line as before in order to make things clear.

**Ex. 10:** $(6x^4 + 13x^3 + 39x^2 + 37x + 45) \div (x^2 - 2x - 9)$

|            |              | $x^4$ | $x^3$ | $x^2$ | $x$ | 1 |
|------------|--------------|-------|-------|-------|-----|-----|
| Divisor    | $x^2 - 2x - 9$ | 6 | 13 | 39 | 37 | 45 |
| Complement | $2 + 9$      |       | 12 | 54 |     |    |
|            |              |       |    | 50 | 225 |    |
|            |              |       |    |    | 286 | 1287 |
| Quotient   |              | $6 \div 1 = 6$ | 25 | 143 | R = 548 | 1332 |

Therefore, Q = $6x^2 + 25x + 143$; R = $548x + 1332$

**Ex. 11:** $(2x^4 - 3x^3 - 3x - 2) \div (x^2 + 1)$

Here we find that the $x^2$ term is missing in the dividend. We must introduce an $x^2$ term with a 0 coefficient. Similarly, we must introduce an $x$ term with a 0 coefficient in the divisor as well. The rest of the steps remain the same.

|  |  | $x^4$ | $x^3$ | $x^2$ | $x$ | 1 |
|---|---|---|---|---|---|---|
| Divisor | $x^2 - 0x + 1$ | 2 | $-3$ | 0 | $-3$ | $-2$ |
| Complement | $0 - 1$ |  | 0 | $-2$ |  |  |
|  |  |  |  | 0 | 3 |  |
|  |  |  |  |  | 0 | 2 |
| Quotient |  | $2 \div 1 = 2$ | $-3$ | $-2$ | $R = 0$ | 0 |

Both terms in the remainder have a coefficient of 0. This means that this division process does not leave a remainder at all.
Therefore, $Q = 2x^2 - 3x - 2$; $R = 0$

**Ex. 12:** $(x^4 - 3x^2 + 3x - 1) \div (x^2 - 2x + 1)$

|  |  | $x^3$ | $x^2$ | $x$ | 1 |
|---|---|---|---|---|---|
| Divisor | $x^2 - 2x + 1$ | 1 | $-3$ | 3 | $-1$ |
| Complement | $2 - 1$ |  | 2 | $-1$ |  |
|  |  |  |  | $-2$ | 1 |
|  |  |  |  | 0 | 0 |
| Quotient |  | $1 \div 1 = 1$ | $-1$ | $R = 0$ | 0 |

Therefore, $Q = x - 1$; $R = 0$

**Ex. 13:** $(2x^3 + 9x^2 + 18x + 20) \div (x^2 + 2x + 4)$

|  |  | $x^3$ | $x^2$ | $x$ | 1 |
|---|---|---|---|---|---|
| Divisor | $x^2 + 2x + 4$ | 2 | 9 | 18 | 20 |
| Complement | $-2 - 4$ |  | $-4$ | $-8$ |  |
|  |  |  |  | $-10$ | $-20$ |
| Quotient |  | $2 \div 1 = 2$ | 5 | $R = 0$ | 0 |

Therefore, $Q = 2x + 5$; $R = 0$

We will now consider the case when the coefficient of the first term of the divisor is not equal to 1.

**Ex.** 14: $(3x^2 - x - 5) \div (3x - 7)$

If we divide the numerator and denominator by the same non-zero number or expression, the resulting quotient remains unaffected. This means we must first take steps to make the coefficient of the first term of the divisor equal to 1. In other words, we divide the divisor by the coefficient of the first term.

Therefore, $(3x - 7) \div 3 = x - 7/3$.

Now that the coefficient of the first term of the divisor is 1, we can use the same procedure as we have done so far.

We must remember to divide the quotient by 3 as well to get the final answer. This is the only additional step in the process.

|  |  | $x^2$ | $x$ | 1 |
|---|---|---|---|---|
| Divisor | $3x - 7$ | 3 | $-1$ | $-5$ |
| Divisor $\div$ 3 | $x - \dfrac{7}{3}$ |  |  |  |
| Complement | $\dfrac{7}{3}$ |  | 7 | 14 |
| Quotient |  | $3 \div 1 = 3$ | 6 | $R = 9$ |

Therefore, our intermediate quotient $= 3x + 6$

Divide the quotient by 3 as well, just as you divided the divisor for getting the final quotient $= (3x + 6) \div 3 = x + 2$

And the Remainder $= 9$

**Ex.** 15: $(2x^5 - 9x^4 + 5x^3 + 16x^2 - 16x + 36) \div (2x^2 - 3x + 1)$

| | | $x^5$ | $x^4$ | $x^3$ | $x^2$ | $x$ | 1 |
|---|---|---|---|---|---|---|---|
| Divisor | $2x^2 - 3x + 1$ | 2 | −9 | 5 | 16 | −16 | 36 |
| Divisor ÷ 2 | $x^2 - \dfrac{3x}{2} + \dfrac{1}{2}$ | | | | | | |
| Complement | $\dfrac{3}{2} - \dfrac{1}{2}$ | | 3 | −1 | | | |
| | | | | −9 | 3 | | |
| | | | | | $\dfrac{-15}{2}$ | $\dfrac{5}{2}$ | |
| | | | | | | $\dfrac{69}{4}$ | $\dfrac{-23}{4}$ |
| Quotient | | $2 \div 1 = 2$ | −6 | −5 | $\dfrac{23}{2}$ | $R = \dfrac{15}{4}$ | $\dfrac{121}{4}$ |

Intermediate Quotient = $2x^3 - 6x^2 - 5x + \dfrac{23}{2}$

Final Quotient = Intermediate Quotient ÷ 2 = $x^3 - 3x^2 - \dfrac{5x}{2} + \dfrac{23}{4}$

Remainder = $\dfrac{15x}{4} + \dfrac{121}{4}$

Some more observations are in order.

In order to get the final quotient, we must divide the intermediate quotient by the same number that we divided the divisor with in order to get the coefficient of the first term to 1. Since we divided the divisor by a number, we must divide the quotient by the same number as well.

1.  In each of these cases, we leave the remainder untouched.
2.  The most exciting observation we can make is this. The method for division that we have used so far is applicable for division of all algebraic expressions or polynomials in $x$.
    a.  Each polynomial in $x$ is simply a decimal number, when $x$ equals 10.
    b.  For example, $x^2 + 2x + 1$ is simply equal to 121 when $x=10$. Therefore, we can use this technique for handling numbers as well.

    i.  For example, the above procedure can be used for solving: $(x^2 + 2x + 1) \div (x + 1)$

    ii.  Therefore, we should be able to use the same procedure for solving $121 \div 11$ which is the same as $(x^2 + 2x + 1) \div (x + 1)$, when $x=10$.

This leads us to the technique for speed division of numbers with larger divisors.

## 12.1.1 Exercises for practice

**Prob 1:** $(x^3 - 3x^2 + 10x - 7) \div (x - 5)$
**Prob 2:** $(x^4 + x^2 + 1) \div (x^2 - x + 1)$
**Prob 3:** $(x^3 + 7x^2 + 9x + 11) \div (x - 2)$
**Prob 4:** $(-2x^5 - 7x^4 + 2x^3 + 18x^2 - 3x - 8) \div (x^3 - 2x^2 + 1)$
**Prob 5:** $(-4x^3 - 7x^2 + 9x - 12) \div (2x - 4)$

In summary, this chapter sets the stage for deploying this technique for handing division. The key learning in this chapter has been the equivalence between a polynomial in $x$, when $x=10$ and a number. The difference in these two worlds is simply the following. The largest number that any place can hold in arithmetic is 9. Any number greater than 9 simply results in a carry forward. Similarly, any number lesser than 0 results in a borrow operation. However, in algebra, the coefficient can be any quantity. The concept of a borrow operation or carry forward is not applicable in the domain of algebra.

Therefore, a polynomial expression like $12x^2 + 13x + 14$ is the same as $_12_13_14$. We resolve the carry forwards and conclude that $_12_13_14 = 1344$. All things fall into place one by one.

## 12.2 First Steps in Speed Division

Now, we will extend the concept we looked at in the previous section. We will apply the concept we learnt in algebraic division to arithmetic division examples.

**Ex. 1:** $111 \div 89$

| 89) | 1 | 1 | 1 | |
|---|---|---|---|---|
| 11 | | | | 100s complement of 89 |
| | | 1 | 1 | Partial quotient × complement |
| = | 1 | 2 | 2 | |

Therefore, Quotient = 1 and Remainder = 22.

**Ex. 2:** $111 \div 73$

| 73) | 1 | 1 | 1 | |
|---|---|---|---|---|
| 27 | | | | 100s complement of divisor |
| | | 2 | 7 | Partial quotient × complement |
| = | 1 | 3 | 8 | |

Therefore, Quotient = 1 and Remainder = 38

**Ex. 3:** $1234 \div 888$

| 888) | 1 | 2 | 3 | 4 | |
|---|---|---|---|---|---|
| 112 | | | | | 100s complement of divisor |
| | | 1 | 1 | 2 | Partial quotient × complement |
| = | 1 | 3 | 4 | 6 | |

Therefore, Quotient = 1 and Remainder = 346

**Ex. 4:** $12123 \div 8978$

| 8978) | 1 | 2 | 1 | 2 | 3 | |
|---|---|---|---|---|---|---|
| 1022 | | | | | | Complement of divisor |
| | | 1 | 0 | 2 | 2 | Partial quotient × complement |
| = | 1 | 3 | 1 | 4 | 5 | |

Therefore, Quotient = 1 and Remainder = 3145

**Ex. 5:** 11123 ÷ 7978

```
7978) | 1 | 1  1  2  3 |
2022  |   |            |              Complement of divisor
      |   | 2  0  2  2 |  Partial quotient × complement
      -------------------
   =  | 1 | 3  1  4  5 |
```

Therefore, Quotient = 1 and Remainder = 3145

**Ex. 6:** 10306 ÷ 7988

```
7988) | 1 | 0  3  0  6 |
2012  |   |            |              10000s complement of divisor
      |   | 2  0  1  2 |  Partial quotient × complement
      -------------------
   =  | 1 | 2  3  1  8 |
```

Therefore, Quotient = 1 and Remainder = 2318

**Ex. 7:** 1010110 ÷ 899999

```
899999) | 1 | 0  1  0  1  1  0 |
100001  |   |                  |              Complement of divisor
        |   | 1  0  0  0  0  1 |  Partial quotient × complement
        ---------------------------
     =  | 1 | 1  1  0  1  1  1 |
```

Therefore, Quotient = 1 and Remainder = 110111

**Ex. 8:** 1115 ÷ 88

```
88) | 1  1 | 1  5 |
12  |      |      |            Complement of divisor
    |    1 | 2    |  Partial quotient × complement
    |      | 2  4 |  Partial quotient × complement
    -----------------
 =  | 1  2 | 5  9 |
```

Therefore, Quotient = 12 and Remainder = 59

**Ex. 9:** 10113 ÷ 91

| 91) | 1 | 0 | 1 | 1 | 3 | |
|---|---|---|---|---|---|---|
| 09 | | 0 | 9 | | | Complement of divisor |
| | | | 0 | 0 | | Partial quotient × complement |
| | | | | 9 | 0 | Partial quotient × complement |
| | | | | | | Partial quotient × complement |
| = | 1 | 0 | ₁0 | ₁0 | 3 | |

Therefore, Quotient = 110 and Remainder = 103

Remainder is greater than the divisor. Therefore, we subtract the divisor from the remainder, 103 − 91 = 12. The final remainder is equal to 12. We increment quotient by 1. The final quotient is 111.

## 12.2.1 Exercises for practice

**Prob 1:** 12354 ÷ 879
**Prob 2:** 210021 ÷ 8999
**Prob 3:** 300001 ÷ 8997
**Prob 4:** 111021 ÷ 8887
**Prob 5:** 200165 ÷ 8988
**Prob 6:** 3003003 ÷ 79999
**Prob 7:** 1030007 ÷ 98997
**Prob 8:** 1111111 ÷ 99989
**Prob 9:** 11201 ÷ 86
**Prob 10:** 123456 ÷ 799

We consider large numbers that are closer to the base. This gives us the complements with small digits. This in turn makes the multiplication process simple and straightforward.

Let us consider 1011 ÷ 23 and see how this works with this method. The divisor is far away from its base, so we need to be ready for bigger products and long winding convergence. Let us actually see what we mean by this.

**Ex. 1:** $1011 \div 23$

| 23) | 1 | 0 | 1 | 1 | |
|---|---|---|---|---|---|
| 77 | | | | | Complement of divisor |
| | 7 | 7 | | | Partial quotient × complement |
| | | $_4$9 | $_4$9 | | Partial quotient × complement |
| = | 1 | 7 | $_5$7 | $_5$0 | |

Therefore, Quotient = 17 and Remainder = 620

The remainder is very large. Whenever the divisor is far away from the base, its complement is large. This produces a cascading effect of resulting in large products and remainders. This forces us to keep dividing the remainder repeatedly. This makes the convergence rather cumbersome and slow.

Instead, if we multiplied 23×4, we get 92. We can solve for $1011 \div 92$ and multiply the quotient by 4. A divisor like 92 gives us a complement of 8 and life becomes immensely simple. Let us use this technique and see how the solution space looks like.

| 92) | 1 | 0 | 1 | 1 | |
|---|---|---|---|---|---|
| 08 | | | | | Complement of divisor |
| | | 0 | 8 | | Partial quotient × complement |
| | | | 0 | 0 | Partial quotient × complement |
| = | 1 | 0 | 9 | 1 | |

Therefore, Intermediate Quotient = 10 and Remainder = 91

Therefore, our final quotient = 10 × 4 = 40 (remember, we multiplied 23 × 4 to get to our temporary divisor of 92; therefore, we need to multiply our intermediate quotient by the same factor).

Remainder is greater than divisor. Therefore, final remainder = 91 − 3×23 = 22; and final quotient = 40 + 3 = 43.

The idea is to look for ways to make subtle changes to the problem to suit our speed division techniques. Let us extend this concept a bit further and applying the algebraic division process more directly. This will produce the next variation in division.

## 12.3 Next Steps in Speed Division

Now, let us take a look at complements a little differently. The 100s complement of 89 is 11. This is because, $89 = x^2 - x - 1$ when $x = 10$. And complement of $x^2 - x - 1 = x + 1$. Therefore we will write this as 1 + 1. In other words, 11 is represented as 1 + 1. The Tens and Units place are separated by a + or a − sign as applicable. This is similar to what we did in the case of algebraic division.

Similarly, the 100s complement of 112 is −1 −2. The 100s complement of 160 is −6 −0. And so on.

Let us quickly write down the complements of the following numbers. This will be critical in handling the speed division technique that we will be looking at.

### 12.3.1 Exercises for practice

Find the complements of the following numbers:
**Prob 1:** 131
**Prob 2:** 87
**Prob 3:** 1234
**Prob 4:** 899
**Prob 5:** 12
**Prob 6:** 93
**Prob 7:** 124
**Prob 8:** 1121
**Prob 9:** 989
**Prob 10:** 1008

Let us now solve a few problems to gain an understanding of this speed division technique that relies on the procedure that we used in algebraic division at the beginning of this chapter.

**Ex. 1:** $1234 \div 112$

| 1 | 1 | 2) | 1 | 2 | 3 | 4 |
|---|---|---|---|---|---|---|
|   | −1 | −2 |   | −1 | −2 |   |
|   |   |   |   |   | −1 | −2 |
| = |   |   | 1 | 1 | 0 | 2 |

Therefore, Quotient = 11 and Remainder = 2

**Ex. 2:** $1341 \div 121$

| 1 | 2 | 1) | 1 | 3 | 4 | 1 |
|---|---|---|---|---|---|---|
|   | −2 | −1 |   | −2 | −1 |   |
|   |   |   |   |   | −2 | −1 |
| = |   |   | 1 | 1 | 1 | 0 |

Therefore, Quotient = 11 and Remainder = 10

**Ex. 3:** $1224 \div 160$

| 1 | 6 | 0) | 1 | 2 | 2 | 4 |
|---|---|---|---|---|---|---|
|   | −6 | 0 |   | −6 | 0 |   |
|   |   |   |   |   | 24 | 0 |
| = |   |   | 1 | −4 | $_2$6 | 4 |

Therefore, Quotient = 6 and Remainder = 264
Remainder is greater than the divisor. We subtract 160 from the remainder and increment the quotient by 1.

We get Quotient = 7 and Remainder = 104

**Ex. 4:** $239479 \div 11203$

| 1 | 1 | 2 | 0 | 3) | 2 | 3 | 9 | 4 | 7 | 9 |
|---|---|---|---|---|---|---|---|---|---|---|
|   | −1 | −2 | 0 | −3 |   | −2 | −4 | 0 | −6 |   |
|   |   |   |   |   |   |   | −1 | −2 | 0 | −3 |
| = |   |   |   |   | 2 | 1 | 4 | 2 | 1 | 6 |

Therefore, Quotient = 21 and Remainder = 4216

**Ex.** 5: $13045 \div 113$

| | | | | | | | | |
|---|---|---|---|---|---|---|---|---|
| 1 | 1 | 3) | 1 | 3 | 0 | 4 | 5 |
| | −1 | −3 | | −1 | −3 | | |
| | | | | | −2 | −6 | |
| | | | | | | 5 | 15 |
| = | | | 1 | 2 | −5 | 3 | $_2$0 |

Therefore, Quotient = 115 and Remainder = 50

**Ex.** 6: $13456 \div 1123$

| | | | | | | | | |
|---|---|---|---|---|---|---|---|---|
| 1 | 1 | 2 | 3) | 1 | 3 | 4 | 5 | 6 |
| | −1 | −2 | −3 | | −1 | −2 | −3 | |
| | | | | | | −2 | −4 | −6 |
| = | | | | 1 | 2 | 0 | −2 | 0 |

Therefore, Quotient = 12 and Remainder = −20

Remainder is negative. Therefore, decrement quotient by one and add the divisor back to the remainder. This gives us a quotient of 11 and remainder of 1103.

**Ex.** 7: $13905 \div 113$

| | | | | | | | | |
|---|---|---|---|---|---|---|---|---|
| 1 | 1 | 3) | 1 | 3 | 9 | 0 | 5 |
| | −1 | −3 | | −1 | −3 | | |
| | | | | | −2 | −6 | |
| | | | | | | −4 | −12 |
| = | | | 1 | 2 | 4 | $-_1$0 | −7 |

Therefore, Quotient = 124 and Remainder = −107

Remainder is negative. Therefore, decrement quotient by one and add the divisor back to the remainder. This gives us a quotient of 123 and remainder of 6.

**Ex. 8:** $11111 \div 1012$

```
    1  0  1  2) | 1  1 ┊ 1   1   1
       0 -1 -2  |    0 ┊-1  -2
               |       ┊ 0  -1  -2
    _____|_____
       =        | 1  1 ┊ 0  -2  -1
```

Therefore, Quotient = 11 and Remainder = –21

Remainder is negative. Therefore, decrement quotient by one and add the divisor back to the remainder. This gives us a quotient of 10 and remainder of 991.

**Ex. 9:** $13999 \div 112$

```
    1  1  2) | 1   3   9 ┊ 9    9
      -1 -2  |    -1  -2 ┊
            |        -2 ┊-4
            |            ┊-5  -10
    _____|_____
      =      | 1   2   5 ┊ 0   -1
```

Therefore, Quotient = 125 and Remainder = –1

Remainder is negative. Therefore, decrement quotient by one and add the divisor back to the remainder. This gives us a quotient of 124 and remainder of 111.

**Ex. 10:** $11329 \div 1132$

```
    1   1   3   2) | 1   1 ┊ 3   2   9
       -1  -3  -2  |    -1 ┊-3  -2
                  |       ┊ 0   0   0
    _____|_____
       =           | 1   0 ┊ 0   0   9
```

Therefore, Quotient = 10 and Remainder = 9

## 12.3.2 Exercises for practice

**Prob 1:** $103 \div 82$
Prob 2: $2341 \div 181$
Prob 3: $39999 \div 9819$

Prob 4: 1111 ÷ 839
Prob 5: 13045 ÷ 988
Prob 6: 5012 ÷ 818
Prob 7: 7111 ÷ 858
Prob 8: 43999 ÷ 828
Prob 9: 1771 ÷ 838
Prob 10: 39893 ÷ 829

Let us now close this discussion with a variation of the problem solving process discussed so far. Let us consider the following example to highlight the variation.

**Ex. 11:** 1699 ÷ 223

The 100s complement of 223 is 123; which is greater than the base itself. The 1000s complement of this number is 777 or −7 −7 −7. Such a huge complement will make the problem solving process very cumbersome. So, this is what we will do.

We will divide the divisor 223 by 2. This gives us 111 ½. The complement of 111 ½ is −1 −1 ½

| 2 | 2 | 3) | 1 | 6 | 9 | 9 | |
|---|---|---|---|---|---|---|---|
| 1 | 1 | $1\frac{1}{2}$ | | | | | Divide the divisor by 2 |
| | −1 | $-1\frac{1}{2}$ | | | | | Complement of divisor |
| | | | −1 | $-1\frac{1}{2}$ | | | |
| | | | | −5 | $-7\frac{1}{2}$ | | |
| = | | | 1 | 5 | $2\frac{1}{2}$ | $1\frac{1}{2}$ | |
| | | | $7\frac{1}{2}$ | $2\frac{1}{2}$ | $1\frac{1}{2}$ | | Divide the interm. quotient by 2 |

Now, the quotient must always be a whole number. So we add the ½ from 7 ½ to the remainder. This is half of 223 (divisor). Therefore the final remainder is 111 ½ + 2 ½ * 10 + 1 ½ = 138.

Therefore, Quotient = 7 and Remainder = 138

Let us consider another problem to highlight this procedure.

**Ex. 12:** 2699 ÷ 224

| 2 | 2 | 4) | 2 | 6 ¦ | 9 | 9 | |
|---|---|----|---|-----|---|---|---|
| 1 | 1 | 2 | | | | | Divide the divisor by 2 |
| | −1 | −2 | | | | | Complement of divisor |
| | | | −2 ¦ | −4 | | | |
| | | | | ¦ −4 | −8 | | |
| = | | | 2 | 4 ¦ | 1 | 1 | |
| | | | 1 | 2 ¦ | 1 | 1 | Divide the interm. quotient by 2 |

Therefore, Quotient = 12 and Remainder = 11

In this chapter, we have explored several techniques that are closely linked to the method of handling algebraic division. Making such linkages between various techniques and understanding the boundary conditions help us to understand and appreciate numerical manipulations better.

# 13 Argumental Division

This is the method, where we ask ourselves, what must be the quotient that satisfies the relationship:

Quotient × divisor + remainder = dividend

This question – "what must" is the argument that we put forward to solve a division challenge. Let us look at a few examples to demonstrate what we mean by argumental[2] division.

**Ex. 1:** $(x^3 + 7x^2 + 9x + 11) \div (x - 2)$

$x^3$ and $x$ are the first terms of the dividend and divisor respectively. The question we ask is what must be multiplied by $x$ in order to get $x^3$ ?
The answer is $x^2$. This is the first term of our quotient.

Quotient: $x^2$

Now, that we have $x^2$ in the quotient, multiplying it with –2 of the divisor gets us $-2x^2$. What must be added to $-2x^2$ in order to get the second term of the dividend, which is $7x^2$ ?
The answer is $9x^2$. Therefore the second term must be $+9x$ which when multiplied by $x$ of the divisor will give us $9x^2$.

Quotient: $x^2 + 9x$

Finally, when –2 of divisor is multiplied with $+9x$ of the quotient, we have $-18x$. What must be added to $-18x$ to get $9x$ in the dividend?
The answer is $+27x$. Therefore the third term of the quotient must be $+27$, which when multiplied by $x$ of the divisor will give $27x$.

Quotient: $x^2 + 9x + 27$

---

[2] We must confess that we were unable to find the word "argumental" in the dictionary we used. We decided to stick with it, because the word does communicate what we are trying to do. You may ignore this term, in case it is confusing to you.

---

Now, we turn our attention to the remainder. The product of −2 of the divisor and 27 of the quotient is −54. What must be added to −54 to get 11 that we see in the dividend?
The answer is 65. Therefore, the remainder is 65.

$$Q = x^2 + 9x + 27\,;\, R = 65$$

**Ex. 2:** $(x^3 - x^2 + 7x + 3) \div (x - 3)$
We would like to use the exact language that we used for the previous example, so things are clear to us.

$x^3$ and $x$ are the first terms of the dividend and divisor respectively. The question we ask is what must be multiplied by $x$ in order to get $x^3$?
The answer is $x^2$. This is the first term of our quotient.

Quotient: $x^2$

Now, that we have $x^2$ in the quotient, multiplying it with −3 of the divisor gets us $-3x^2$. What must be added to $-3x^2$ in order to get the second term of the dividend, $-x^2$?
The answer is $2x^2$. Therefore the second term must be $+2x$ which when multiplied by $x$ of the divisor will give us $2x^2$.

Quotient: $x^2 + 2x$

Finally, when −3 of the divisor is multiplied with $+2x$ of the quotient, we have $-6x$. What must be added to $-6x$ to get $7x$ in the dividend?
The answer is $+13x$. Therefore the third term of the quotient must be $+13$, which when multiplied by $x$ of the divisor will give $13x$.

Quotient: $x^2 + 2x + 13$

Now, we turn our attention to determine the remainder. The product of −3 of the divisor and $+13$ of the quotient is −39. What must be added to −39 to get $+3$ that we see in the dividend?
The answer is 42. Therefore, the remainder is 42.

$Q = x^2 + 2x + 13 ; R = 42$

**Ex. 3:** $(7x^2 + 5x + 3) \div (x - 1)$

$7x^2$ and $x$ are the first terms of the dividend and divisor respectively. The question we ask is what must be multiplied by $x$ in order to get $7x^2$ ? The answer is $7x$. This is the first term of our quotient.

Quotient: $7x$

Now that we have $7x$ in the quotient, multiplying it with $-1$ of the divisor gets us $-7x$. What must be added to $-7x$ in order to get the second term of the dividend, $+5x$? The answer is $12x$. Therefore the second term must be $+12$ which when multiplied by $x$ of the divisor will give us 12.

Quotient: $7x + 12$

Now, we turn our attention to determine the remainder. The product of $-1$ of divisor and $+12$ of the quotient is $-12$. What must be added to $-12$ to get $+3$ that we see in the dividend? The answer is $+15$. Therefore, the remainder is 15.

$Q = 7x + 12 ; R = 15$

**Ex. 4:** $(7x^2 + 5x + 3) \div (x + 1)$

$7x^2$ and $x$ are the first terms of the dividend and divisor respectively. The question we ask is what must be multiplied by $x$ in order to get $7x^2$ ? The answer is $7x$. This is the first term of our quotient.

Quotient: $7x$

Now, that we have $7x$ in the quotient, multiplying it with $+1$ of the divisor gets us $+7x$. What must be added to $+7x$ in order to get the second term of the dividend $+5x$? The answer is $-2x$. Therefore the second term must be $-2$ which when multiplied by $x$ of the divisor will give us $-2x$.

Quotient: $7x - 2$

Now, we turn our attention to determine the remainder. The product of 1 of divisor and –2 of the quotient is –2. What must be added to –2 to get +3 that we see in the dividend? The answer is +5. Therefore, the remainder is 5.

$Q = 7x - 2 \, ; R = 5$

**Ex. 5:** $(3x^2 - x - 5) \div (3x - 7)$

$3x^2$ and $x$ are the first terms of the dividend and divisor respectively. The question we ask is what must be multiplied by $3x$ in order to get $3x^2$? The answer is $x$. This is the first term of our quotient.

Quotient: $x$

Now, that we have $x$ in the quotient, multiplying it with –7 of the divisor gets us $-7x$. What must be added to $-7x$ in order to get the second term of the dividend $-x$? The answer is $6x$. Therefore the second term must be +2 which when multiplied by $3x$ of the divisor will give us $6x$.

Quotient: $x + 2$

Now, we turn our attention to determine the remainder. The product of –7 of divisor and +2 of the quotient is –14. What must be added to –14 to get –5 that we see in the dividend? The answer is +9. Therefore, the remainder is 9.

$Q = x + 2 \, ; R = 9$

**Ex. 6:** $(16x^2 + 8x + 1) \div (4x + 1)$

$16x^2$ and $4x$ are the first terms of the dividend and divisor respectively. The question we ask is what must be multiplied by $4x$ in order to get $16x^2$?
The answer is $4x$. This is the first term of our quotient.

Quotient: $4x$
Now, that we have $4x$ in the quotient, multiplying it with +1 of the divisor gets us $4x$. What must be added to $4x$ in order to get the second term of the dividend $8x$?

The answer is 4x. Therefore the second term must be +1 which when multiplied by 4x of the divisor will give us 4x.

Quotient: $4x + 1$

Now, we turn our attention to determine the remainder. The product of +1 of divisor and +1 of the quotient is +1. What must be added to +1 to get +1 that we see in the dividend? The answer is 0. Therefore, the remainder is 0.

$Q = 4x + 1; R = 0$

We have used the same or similar argument to handle divisors, which have 3 terms. This shows that the procedure is generic enough to be applied to a variety of expressions.

**Ex. 7:** $(x^3 + 2x^2 + 3x + 5) \div (x^2 - x - 1)$

$x^3$ and $x^2$ are the first terms of the dividend and divisor respectively. The question we ask is what must be multiplied by $x^2$ in order to get $x^3$? The answer is $x$. This is the first term of our quotient.

Quotient: $x$

Now, that we have $x$ in the quotient, multiplying it with $-x$ of the divisor gets us $-x^2$. What must be added to $-x^2$ in order to get the second term of the dividend $2x^2$?
The answer is $3x^2$. Therefore the second term must be +3 which when multiplied by $x^2$ of the divisor will give us $3x^2$.

Quotient: $x + 3$

Now we turn our attention to the task of determining the remainder. Here is a thumb rule. As soon as we complete the determination of the constant term of the quotient, we move to working on the remainder.
Finally, when +3 of divisor is multiplied with $-x$ of the quotient, we have $-3x$. And the product of $-1$ of the divisor and $x$ in the coefficient gives us $-x$. So we have $-3x + (-x) = -4x$ in all. What must be added to $-4x$ to

get $3x$ in the dividend? The answer is $+7x$. Therefore the first term of the remainder must be $+7x$.

Quotient: $x + 3$; R $= 7x$

The product of -1 of divisor and +3 of the quotient is $-3$. What must be added to $-3$ to get $+5$ that we see in the dividend? The answer is 8. Therefore, the second term of the remainder is 8..

Quotient: $x + 3$; R $= 7x + 8$

**Ex. 8:** $(x^4 + 4x^3 + 6x^2 + 4x + 1) \div (x^2 + 2x + 1)$

$x^4$ and $x^2$ are the first terms of the dividend and divisor respectively. The question we ask is what must be multiplied by $x^2$ in order to get $x^4$? The answer is $x^2$. This is the first term of our quotient.

Quotient: $x^2$

Now, that we have $x^2$ in the quotient, multiplying it with $2x$ of the divisor gets us $2x^3$. What must be added to $2x^3$ in order to get the second term of the dividend $4x^3$? The answer is $2x^3$.
Therefore the second term must be $+2x$ which when multiplied by $x$ of the divisor will give us $2x^3$.

Quotient: $x^2 + 2x$

Now let us consider the cross products of terms in the divisor and quotient that contribute to $x^2$
- 1 in the divisor multiplied by $x^2$ in the quotient gets us $x^2$
- $2x$ in the divisor and $2x$ in the quotient gets us $4x^2$
- We have $4x^2 + x^2 = 5x^2$

What must be added to $5x^2$ in order to get $6x^2$ that we see in the dividend?
The answer is $x^2$. Therefore, to get $x^2$, we must multiply the $x^2$ in the divisor with 1.

Quotient: $x^2 + 2x + 1$

Now, we turn our attention to determine the remainder. The product of +1 of divisor and +1 of the quotient is +1. What must be added to +1 to get +1 that we see in the dividend? The answer is 0. Therefore, the remainder is 0.

$Q = x^2 + 2x + 1$; $R = 0$

**Ex. 9:** $(x^4 + 2x^3 + 3x^2 + 2x + 1) \div (x^2 + x + 1)$

$x^4$ and $x^2$ are the first terms of the dividend and divisor respectively. The question we ask is what must be multiplied by $x^2$ in order to get $x^4$? The answer is $x^2$. This is the first term of our quotient.

Quotient: $x^2$

Now, that we have $x^2$ in the quotient, multiplying it with $x$ of the divisor gets us $x^3$. What must be added to $x^3$ in order to get the second term of the dividend $2x^3$?
The answer is $x^3$. Therefore the second term must be $+x$ which when multiplied by $x^2$ of the divisor will give us $2x^3$.

Quotient: $x^2 + x$

Now let us consider the cross products of terms in the divisor and quotient that contribute to $x^2$.

- 1 in the divisor multiplied by $x^2$ in the quotient gets us $x^2$
- $x$ in the divisor and $x$ in the quotient gets us $x^2$
- We have $x^2 + x^2 = 2x^2$

What must be added to $2x^2$ in order to get $3x^2$ that we see in the dividend?
The answer is $x^2$. Therefore, to get $x^2$, we must multiply the $x^2$ in the divisor with 1.

Quotient: $x^2 + x + 1$

Now, we turn our attention to determine the remainder. The product of $+1$ of divisor and $+1$ of the quotient is $+1$. What must be added to $+1$ to get $+1$ that we see in the dividend? The answer is 0. Therefore, the remainder is 0.

$Q = x^2 + x + 1; R = 0$

**Ex. 10:** $(12x^4 - 3x^3 - 3x - 12) \div (x^2 + 1)$

Let us introduce a 0 coefficient to missing powers of $x$ in the dividend and divisor. We have: $(12x^4 - 3x^3 - 0x^2 - 3x - 12) \div (x^2 + 0x + 1)$

$12x^4$ and $x^2$ are the first terms of the dividend and divisor respectively. The question we ask is what must be multiplied by $x^2$ in order to get $12x^4$ ?

The answer is $12x^2$. This is the first term of our quotient.

Quotient: $12x^2$

Now, that we have $x^2$ in the quotient, multiplying it with $x$ of the divisor gets us 0. What must be added to 0 in order to get the second term of the dividend $-3x^3$ ?

The answer is $-3x^3$. Therefore the second term must be $-3x$ which when multiplied by $x^2$ of the divisor will give us $-3x^3$.

Quotient: $12x^2 - 3x$

Now let us consider the cross products of terms in the divisor and quotient that contribute to $x^2$

- 1 in the divisor multiplied by $12x^2$ in the quotient gets us $12x^2$
- $0x$ in the divisor and $-3x$ in the quotient gets us 0
- We have $12x^2 + 0 = 12x^2$ as the co-efficient

What must be added to $12x^2$ in order to get $0x^2$ that we see in the dividend?

The answer is $-12x^2$. Therefore, to get $0x^2$, we must multiply the $x^2$ in the divisor with $-12$.

Quotient: $12x^2 - 3x - 12$

Now, we turn our attention to determine the remainder. The product of $+1$ of divisor and $-12$ of the quotient is $-12$. What must be added to $-12$ to get $-12$ that we see in the dividend? The answer is $0$. Therefore, the remainder is $0$.

$Q = 12x^2 - 3x - 12$ ; R = 0

## 13.1.1 Exercises for practice

The following problems will help you get better at this method of division:

**Prob. 1:** $(x^4 - 4x^2 + 12x - 9) \div (x^2 + 3x - 3)$

**Prob. 2:** $(2x^3 + 9x^2 + 18x + 20) \div (2x + 5)$

**Prob. 3:** $(12x^4 + 41x^3 + 81x^2 + 79x + 42) \div (4x^2 + 7x + 6)$

**Prob. 4:** $(2x^3 + 9x^2 + 18x + 20) \div (x^2 + 2x + 4)$

**Prob. 5:** $(x^4 + 3x^3 - 16x^2 + 3x + 1) \div (x^2 + 6x + 1)$

# 14 Straight Division

In this chapter, we will figure out yet another way to approach problems in division. In order to do this, we will solve a few examples and identify the pattern and process for problem solving. You will soon see that the basic approach is reasonably similar to the usual division process we are taught in school. We will, of course, lay the game board a little differently. And you will soon see how this helps us.

## 14.1 Single digit division

**Ex. 1:** $6666 \div 5$

This is an easy one to tackle. The beauty of learning a new technique using an easy example is obvious. We can quickly draw parallels between what we know and create linkages to what we are doing. This helps us internalize the problem solving process.

$$5 \mid 6 \quad 6 \quad 6 \quad 6 \quad \mid$$

*Step 1*: We consider the first digit of the dividend. We perform $6 \div 5$; this gives us a quotient of 1 and remainder 1. We write the quotient below the digit of the dividend under consideration. We will write the remainder in between the first two digits of the dividend as shown below.

| 5 | 6 | 6 | 6 | 6 | |
|---|---|---|---|---|---|
| | | 1 | | | $6 \div 5$ gives R=1 |
| | 1 | | | | $6 \div 5$ gives Q=1 |

*Step 2*: We consider the remainder 1 as the prefix to the second digit in the dividend. We get 16. Now we perform $16 \div 5$. We get a quotient of 3 and remainder equals to 1. We write the quotient below the second digit of the dividend. We write the remainder in between the second and the third digits of the remainder. This is similar to step 1. We always write the quotient below the digit of the dividend under consideration. We write the remainder in the middle of the two digits of dividend as seen in steps 1 and 2.

| 5 | 6 | 6 | 6 | 6 | |
|---|---|---|---|---|---|
| | | 1 | 1 | | 16 ÷ 5 gives R=1 |
| | 1 | 3 | | | 16 ÷ 5 gives Q=3 |

*Step 3*: Continuing with the same process as in previous steps, we read the remainder the previous step as the prefix to the next digit in the dividend. We get 16. We perform 16÷5; we get Q=3 and R=1. We write the results down just like we did in Step 2.

| 5 | 6 | 6 | 6 | 6 | |
|---|---|---|---|---|---|
| | | 1 | 1 | 1 | 16 ÷ 5 gives R=1 |
| | 1 | 3 | 3 | | 16 ÷ 5 gives Q=3 |

*Step 4*: Continuing with the same process as in previous steps, we read the remainder the previous step as the prefix to the next digit in the dividend. We get 16. We perform 16÷5; we get Q=3 and R=1. We write the results down just like we did in Step 3.

| 5 | 6 | 6 | 6 | 6 | | |
|---|---|---|---|---|---|---|
| | | 1 | 1 | 1 | 1 | 16 ÷ 5 gives R=1 |
| | 1 | 3 | 3 | 3 | | 16 ÷ 5 gives Q=3 |

*Step 5*: There are no more digits in the dividend to be considered. Therefore we terminate the process of division.

Therefore, 6666÷5 gives Q=1333 and R=1. Let us solve a few examples to get better at this technique. This technique is similar to the one you have used in your schools. The main difference lies in the way you write down the remainder.

In the classical division technique, we would have solved the problem as below.

```
        | Q:   1   3   3   3
  ------|--------------------
    5   |      6   6   6   6
        |  -   5
        |      1   6
        |  -   1   5
        |          1   6
        |      -   1   5
        |              1   6
        |          -   1   5
        |                  1
```

*Note*: Remainder becomes prefix to the next digit in the dividend. We would have continued doing so, until we ran out of digits in the dividend. This is the similarity that we talked about. It is critical that we look at these patterns and draw linkages between what we know and what we are learning now.

Let us now consider another example.

**Ex. 2:** 5783 ÷ 4

*Step 1*: We start off the process with the first digit of the dividend: 5÷4

| 4 | 5 | 7 | 8 | 3 | |
|---|---|---|---|---|---|
|   |   | 1 |   |   | 5 ÷ 4 gives R=1 |
|   | 1 |   |   |   | 5 ÷ 4 gives Q=1 |

*Step 2*: Remainder 1 becomes prefix to the next digit of the dividend. Now we perform 17÷4; Q=1 and R=1.

| 4 | 5 | 7 | 8 | 3 | |
|---|---|---|---|---|---|
|   |   | 1 | 1 |   | 17 ÷ 4 gives R=1 |
|   | 1 | 4 |   |   | 17 ÷ 4 gives Q=4 |

*Step 3*: Now we perform 18÷4. This results in Q=9 and R=2.

| 4 | 5 | 7 | 8 | 3 | |
|---|---|---|---|---|---|
| | 1 | 1 | 2 | | $18 \div 4$ gives R=2 |
| | 1 | 4 | 4 | | $18 \div 4$ gives Q=4 |

*Step 4*: Similarly, $23 \div 4$ gives Q=5 and R=3.

| 4 | 5 | 7 | 8 | 3 | |
|---|---|---|---|---|---|
| | 1 | 1 | 2 | 3 | $23 \div 4$ gives R=3 |
| | 1 | 4 | 4 | 5 | $23 \div 4$ gives Q=5 |

*Step 5*: We have run out of digits in the dividend. Therefore, $5783 \div 4$ gives Q=1445 and R=3.

Now that we understand the process; let us write the quotient and remainder in one go.

**Ex. 3:** $5367 \div 8$

| 8 | 5 | 3 | 6 | 7 |
|---|---|---|---|---|
| | | 5 | 0 | 7 |
| | 6 | 7 | 0 | |

Therefore the Quotient = 670; Remainder = 7

**Ex. 4:** $8734 \div 7$

| 7 | 8 | 7 | 3 | 4 |
|---|---|---|---|---|
| | | 1 | 3 | 5 | 5 |
| | 1 | 2 | 4 | 7 |

Therefore the Quotient = 1247; Remainder = 5

**Ex. 5:** $8978 \div 6$

| 6 | 8 | 9 | 7 | 8 |
|---|---|---|---|---|
| | | 2 | 5 | 3 | 2 |
| | 1 | 4 | 9 | 6 |

Therefore the Quotient = 1496; Remainder = 2

## 14.1.1 Exercises for practice

**Prob.** 1: 2749 ÷ 6

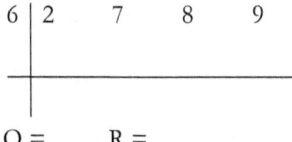

6 | 2    7    8    9

Q =    , R =

**Prob.** 2: 8724 ÷ 7

7 | 8    7    2    4

Q =    , R =

**Prob.** 3: 8121 ÷ 4

4 | 8    1    2    1

Q =    , R =

**Prob.** 4: 1698 ÷ 5

5 | 1    6    9    8

Q =    , R =

**Prob.** 5: 6581 ÷ 8

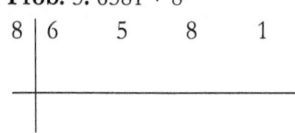

8 | 6    5    8    1

Q =    , R =

**Prob.** 6: 8139 ÷ 5

5 | 8    1    3    9

Q =    , R =

**Prob.** 7: 1199 ÷ 4

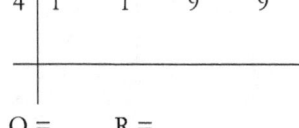

4 | 1    1    9    9

Q =    , R =

**Prob.** 8: 1464 ÷ 4

4 | 1    4    6    4

Q =    , R =

**Prob.** 9: 5625 ÷ 5

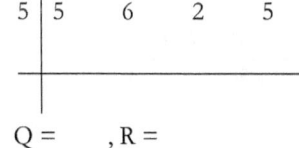

5 | 5    6    2    5

Q =    , R =

**Prob.** 10: 4900 ÷ 8

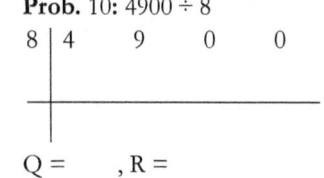

8 | 4    9    0    0

Q =    , R =

## 14.2 Two digit divisions

Now, we will proceed to extend the technique to two digit divisors. Clearly, we will require a few modifications and extensions to the method. While handling two digit divisors, we consider the first digit of the divisor as the base divisor and use the second digit as a flag number. For example, if the divisor is 39; the base divisor will be 3 and the flag will be 9. If the divisor is 87, the base divisor will be 8 and the flag will be 7. The rest of the section deals with how we use the base divisor and flag to complete the division process.

In order to understand the process, let us solve a few problems. And our first problem must be an easy one. This helps us to focus on the technique; while helping us get interesting insights into the solution space.

**Ex. 1:** 971 ÷ 23

$$23 \mid 9 \quad 7 \quad 1$$

We will represent the same problem below, but separating the base divisor and the flag. They have been written in a diagonal arrangement. Please do not read this as 2-cubed. This is still 23.

$$3 \mid 9 \quad 7 \quad 1$$
$$2$$

First observation that we need to make is that we will draw a remainder line; between 7 and 1. This means that number of digits in the flag is also the number of digits in the remainder side of the dividend.

$$3 \mid 9 \quad 7 \quad \vdots \quad 1$$
$$2$$

*Step 1:* Now that we have completed the pre-work; let us proceed to the actual division process. The first digit of the dividend (9) is divided by the base divisor (2).

| 3 | 9 | 7 | ⋮ 1 | |
|---|---|---|---|---|
| 2 | | 1 | ⋮ | 9 ÷ 2 gives R=1 |
| | 4 | | ⋮ | 9 ÷ 2 gives Q=4 |

*Step 2*: The new dividend is 17 (prefix remainder to the next dividend digit). Subtract the product of partial coefficient 4 and flag 3. In other words:
$$17 - 4×3 = 5.$$ Use 5 as the new dividend.

Now, perform $5 ÷ 2$. This gives a quotient of 2 and remainder 1. The 1 comes under the R-line. Draw the R line after the coefficient 2.

| 3 | 9 | 7 | | 1 | |
|---|---|---|---|---|---|
| 2 | | 1 | 1 | | $5 ÷ 2$ gives R=1 |
| | 4 | 2 | | | $5 ÷ 2$ gives Q=2 |

*Step 3*: The new dividend is 11. We subtract the product of 2{partial coefficient} × flag from 11.
$$11 - 2×3 = 5.$$ Use 5 as the new dividend

We have crossed the R-line. Therefore; 5 is the final remainder and quotient is 42.

| 3 | 9 | 7 | | 1 | | |
|---|---|---|---|---|---|---|
| 2 | | 1 | 1 | | 5 | $11 - 2×3 = 5$ |
| | 4 | 2 | | | | |

Therefore Q=42 and R=5

*Note*: The other way of solving this problem is to use argumental division. 971 is $9x^2 + 7x + 1$ and 23 is $2x + 3$ when $x = 10$

Do: $(9x^2 + 7x + 1) ÷ (2x + 3)$ and substitute $x = 10$ in the final answer for quotient and remainder. This is a valid approach as well.

**Ex. 2:** $7458 ÷ 127$

| 7 | 7 | 4 | 5 | | 8 |
|---|---|---|---|---|---|
| 12 | | 14 | 14 | | 92 |
| | 5 | 8 | | | |

Therefore, Quotient = 58; Remainder = 92.

This problem has a nice variation. How did we get a partial reminder of 5 in the first step? Here is the detailed explanation for this.
*Step 1*: $74 ÷ 12$ gives quotient of 6 and remainder of 2.

*Step 2*: We get 25 – 5×7 = -35, which is a negative number. Therefore, we go a step back and reduce the quotient from 6 to 5. This means that we need to add the base divisor to the remainder. Therefore remainder becomes 2+12 = 14. This is how the partial quotient is 5 and remainder is 14. Now 145 – 5×7 = 110, which is the next divisor.

*Step 3*: 110 ÷ 12 gives quotient of 9 and remainder of 2.

*Step 4*: We get 28 - 8×7 again negative. So, we go back a step and reduce the quotient from 9 to 8. The remainder then becomes 2+12=14. Since 148 - 8×7= 92, 92 is the final remainder.

Therefore, Quotient = 58; Remainder = 92.

**Ex. 3:** 948 ÷ 32

| | 2 | 9 | 4 | 8 | |
|---|---|---|---|---|---|
| 3 | | 3 | 3 | | 20 |
| | 2 | 9 | | | |

Therefore, Quotient = 29; Remainder = 20.

**Ex. 4:** 7143 ÷ 1171

| | 71 | 7 | 1 | 4 | 3 |
|---|---|---|---|---|---|
| 11 | | | 5 | | |
| | | 6 | | | |

Quotient = 6, Remainder = 543 – 6×71 = 543 – 426 = 117

**Ex. 5:** 5711 ÷ 54

| | 4 | 5 | 7 | 1 | 1 | |
|---|---|---|---|---|---|---|
| 5 | | 0 | 3 | 6 | | 41 |
| | | 1 | 0 | 5 | | |

Quotient = 105; Remainder = 61 – 20 = 41

## 14.2.1 Summary

For two digit divisors, the digit in the tens place is the base divisor while the digit in the units place becomes the flag.

The number of digits in the flag is also the number of digits in the remainder; draw the R-line off the same number of digits from the right.

Divide the first digits or the required set of digits by the base divisor. Put down the coefficient and the remainder.

The current dividend is the remainder prefixed with the next digit in the dividend. Subtract the product of the quotient and the flag from the current dividend. This difference is our new dividend. Divide the new dividend with the base divisor and repeat the process until you cross the R-line.

Once you cross the R-line, the difference of the current dividend and the product of the quotient and flag is the remainder. We stop the division process.

The quotient is already captured in our working, progressively.

## 14.2.2 Exercises for practice

**Prob. 1:** $3105 \div 67$

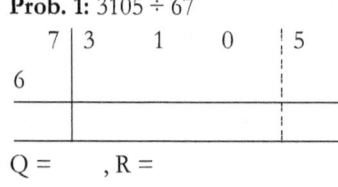

$$Q = \qquad , R =$$

**Prob. 2:** $4922 \div 43$

$$Q = \qquad , R =$$

**Prob. 3:** $6341 \div 72$

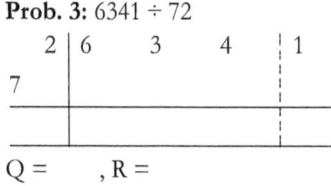

$$Q = \qquad , R =$$

**Prob. 4:** $3617 \div 25$

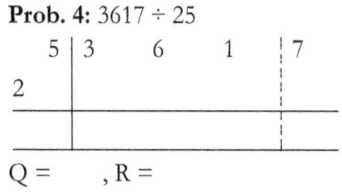

$$Q = \qquad , R =$$

**Prob. 5:** 8816 ÷ 33

```
   3 | 8    8    1  ¦ 6
   3   |                ¦
       |_____¦____
       |                ¦
  _____|_____¦____
       |                ¦
```

Q =      , R =

**Prob. 6:** 4821 ÷ 94

```
        4 | 4    8    2  ¦ 1
        9   |             ¦
            |_____¦____
            |             ¦
       _____|_____¦____
            |             ¦
```

Q =      , R =

**Prob. 7:** 14958 ÷ 48

```
            8 | 1    4    9    5  ¦ 8
              4   |                ¦
            _____|_____¦____
                  |                ¦
            _____|_____¦____
                  |                ¦
```

Q =      , R =

**Prob. 8:** 15799 ÷ 82

```
            2 | 1    5    7    9  ¦ 9
              8   |                ¦
            _____|_____¦____
                  |                ¦
            _____|_____¦____
                  |                ¦
```

Q =      , R =

**Prob. 9:** 83380 ÷ 88

```
            8 | 8    3    3    8  ¦ 0
              8   |                ¦
            _____|_____¦____
                  |                ¦
            _____|_____¦____
                  |                ¦
```

Q =      , R =

**Prob. 10:** 89718 ÷ 79

```
            9 | 8    9    7    1  ¦ 8
              7   |                ¦
            _____|_____¦____
                  |                ¦
            _____|_____¦____
                  |                ¦
```

Q =      , R =

**Prob. 11:** 506907 ÷ 22

```
        2 | 5    0    6    9    0   ¦ 7
    2     |
          |_____
          |
```

Q =        , R =

**Prob. 12:** 620509 ÷ 21

```
        1 | 6    2    0    5    0   ¦ 9
    2     |
          |_____
          |
```

Q =        , R =

**Prob. 13:** 171341 ÷ 91

```
        1 | 1    7    1    3    4   ¦ 1
    9     |
          |_____
          |
```

Q =        , R =

**Prob. 14:** 497113 ÷ 13

```
        3 | 4    9    7    1    1   ¦ 3
    1     |
          |_____
          |
```

Q =        , R =

**Prob. 15:** 883422 ÷ 44

```
        4 | 8    8    3    4    2   ¦ 2
    4     |
          |_____
          |
```

Q =        , R =

**Prob. 16:** 641801 ÷ 83

```
        3 | 6    4    1    8    0   | 1
      8   |                         |
          |-------------------------|--
          |                         |
```

Q =        , R =

**Prob. 17:** 558459 ÷ 35

```
        5 | 5    5    8    4    5   | 9
      3   |                         |
          |-------------------------|--
          |                         |
```

Q =        , R =

**Prob. 18:** 288793 ÷ 39

```
        9 | 2    8    8    7    9   | 3
      3   |                         |
          |-------------------------|--
          |                         |
```

Q =        , R =

**Prob. 19:** 774174 ÷ 17

```
        7 | 7    7    4    1    7   | 4
      1   |                         |
          |-------------------------|--
          |                         |
```

Q =        , R =

**Prob. 20:** 188420 ÷ 13

```
        3 | 1    8    8    4    2   | 0
      1   |                         |
          |-------------------------|--
          |                         |
```

Q =        , R =

# 15 Closing Thoughts

In this book, we have covered several techniques related to speed mathematics. Number patterns and underlying nuances are like a work of art. The more you apply yourself to the subject, the more you uncover and understand. Mathematics like subject which require mastery, is not a spectator sport. This is not something that you relax on your couch, casually browse through and hope to achieve mastery. This will require patience and application.

There are few thumb rules I would like to impress upon our young readers.

1.  Neatness is conducive to accuracy. Refrain from the temptation to write down something quickly and then scratch the same to make the necessary corrections.
2.  One of the weakness we find in student while solving word problems is the usage of = sign. This sign as a specific meaning in the world of mathematics. It cannot be used as a way to begin every new line of step in the problem solving process. Use appropriate mathematical signs and symbols. Never use them to mean something vague. = Sign is never good space filler.
3.  Spend a second or two to explain how you arrived at a certain step. Several books and references use a statement, such as "it follows from the above statement". We have oftentimes wondered how the expression or equation below follows from the one above.
4.  When you are faced with several conclusions during a problem solving process, it is a good idea to number the statements or equations. In subsequent steps, we can refer to this conclusion by using the label or the assigned equation number.
5.  The easiest of problems attracts the silliest of mistakes. If the problem is easy, motivate yourself to get it right. Do not transform this into carelessness.

It has been our sincere effort to gather several techniques from two schools of thought – Vedic and Trachtenberg. The purpose is to present a bouquet of techniques and tricks into the thought arsenal of the young and uninitiated. We have tried to stay out of the debate on which one of these better. It is our belief that each of these techniques has a place in the world of speed math. The curiosity and love for numbers and music in the number patterns has long been an addiction of a kind. If this book succeeds to set a spark of excitement in the reader and compels him or her to explore this universe a bit further, this book would have achieved its objective.

Cheers,
Chandramouli Mahadevan

On behalf of Team Astrarka
August 2010.

www.ingramcontent.com/pod-product-compliance
Lightning Source LLC
Chambersburg PA
CBHW071411170526
45165CB00001B/234